RADON

RISK
AND
REMEDY

RADON

RISK
AND
REMEDY

David J. Brenner, Ph.D.

Radiological Research Laboratory
Columbia University

W. H. FREEMAN AND COMPANY
New York

Library of Congress Cataloging-in-Publication Data

Brenner, David J.
  Radon : Risk and Remedy.

  Includes index.
  1. Radon—Health aspects. 2. Radon—Environmental
aspects. 3. Indoor air pollution—Health aspects.
4. Housing and health. 5. Consumer education. I. Title.
RA1247.R33B74 1989   363.1'79   88-33529
ISBN 0-7167-2029-9
ISBN 0-7167-2030-2 (ppk.)

Printed in the United States of America

1 2 3 4 5 6 7 8 9 0 VB 7 6 5 4 3 2 1 0 8 9

To my parents

Contents

Preface

Avoiding risks is a necessary part of our everyday lives. To avoid the risk of an accident, we take care when crossing the street; to avoid the risk of lung cancer, we might quit or cut down on cigarette smoking. Of course we cannot spend all our time thinking about avoiding risks—if we did, no time would remain to enjoy ourselves! Ideally, then, we would like to reduce our chances of early death or illness without devoting too much time or effort to the problem. So we need to identify the most easily avoidable risks in our lives and then find ways to reduce them.

Radon, a naturally occurring radioactive gas that can build up to high levels in ordinary houses, is one of the more significant of these avoidable hazards. Each year in the United States alone, from 15,000 to 50,000 people may needlessly die from the effects of radon. Those numbers make radon about as dangerous as homicides or road accidents.

Since it is a product of natural forces and only vaguely understood by most homeowners, the radon problem is often thought of as being unavoidable. Part of this book's purpose is to dispel that assumption. While we know that radon is a major killer, we also know how to minimize the dangers it presents. In short, if there is a high level of radon in your home, it can be effectively reduced without too much trouble or expense.

The aim of this book is twofold. First, it outlines the actual dangers of radon. With that background, it moves on to the practical issues: Is your house affected? If it is, what remedies are available? The first step is to test for radon in your home. If high radon levels are discovered, the book explains the relatively small changes that need to be made in your house to fix the problem. Making these alterations and reducing excessive radon levels makes sense in terms of maintaining both your health and the real estate value of your property.

The risks from radon can be managed, but only if we first understand the problem. If awareness of radon's dangers is increased and reasonable precautions are taken, many lives can be saved. If this book goes some way toward that goal, its author will be amply rewarded.

ACKNOWLEDGMENTS

This book greatly profited from the help of many people. In particular, I am grateful to Dr. Eric Hall, Dr. Charles Geard, Dr. Marco Zaider from my own laboratory, the Radiological Research Laboratory. The incisive comments of Dr. Wayne Lowder of the U.S. Department of Energy enlightened me on many occasions.

Thanks are also due to Phil Bourne, Michaela Delegianis, Maria Georgsson, Maria Ruotalo, and Janie Weiss, all of Columbia University, whose continual encouragement led to the completion of this book. Finally, I am happy to acknowledge the help and assistance of Gary Carlson, Director of Acquisitions of W. H. Freeman and Company, and Harold Somers of the University of Manchester.

My own research is supported by the National Cancer Institute and the U.S. Department of Energy.

David J. Brenner
November, 1988

RADON

RISK
AND
REMEDY

Introduction

1

RADON is an invisible, odorless gas that seeps out of rocks and soil into the air. It occurs—and always has occurred—perfectly naturally. Radon gas is also naturally radioactive. It can seep through floors into houses and, if the conditions are right, can build up to very high levels indoors.

Radon is extremely toxic. If radon is inhaled, the result will be radioactive material in the lungs. All radioactive materials can cause cancer, and there is little doubt that breathing radon causes lung cancer. It has been estimated that between 5,000 and 20,000 people die every year in the United States alone because of exposure to radon.

Although some radon is present in all houses, the problem is controllable. If a house has a high level of radon, it can be reduced without too much trouble or expense. But high levels of radon will go undetected unless people make a decision to test their homes. The responsibility for testing firmly rests on the individual.

At the moment, however, most people do not appreciate the extent of the radon problem. For example, a recent New Jersey survey indicated that fewer than one state resident in ten thinks that radon might be a problem in his or her home, and only one in thirty has or is planning to have a home radon test. Compare this finding (see Figure 1-1) with another recent study showing that **one in three** houses in New Jersey probably has a radon problem.

This book is designed to help people with no background in science to become informed about radon. We shall be discussing four basic questions:

• What is radon?

• Just how dangerous is it?

• How can you find out if your house has a problem?

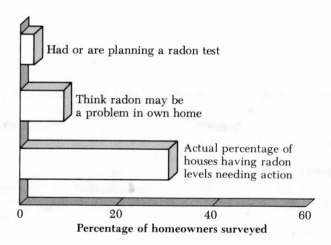

Had or are planning a radon test

Think radon may be a problem in own home

Actual percentage of houses having radon levels needing action

Percentage of homeowners surveyed

Figure 1-1 Comparing people's perception of the radon problem with the extent of the problem itself. The information, for New Jersey, is from the *Star-Ledger*/Eagleton Institute of Politics and from the New Jersey Department of Environmental Protection survey (1987).

- What can you do if your house does have a high level of radon?

Because radon is a naturally occurring gas, the amount of radon leaking out of the ground in 1985 was the same as the amount in 1975 (or 1875, for that matter). Yet only since the mid-1980s has there been much public concern about radon. In an extraordinary incident in December 1984, a construction engineer named Stanley Watras was going to work at the new Limerick nuclear power plant in Pottstown, Pennsylvania. Watras and his family lived in an ordinary suburban house in nearby Boyertown (Figure 1-2). Like all nuclear power plants,

FIGURE 1-2 The Watras family in front of their house in Boyertown, Pennsylvania. In late 1984, their house was found to have radon levels more than 700 times higher than current federal guidelines. (*Newsday*/John H. Cornell, Jr.)

Limerick has radiation detectors at its entrance to prevent people from taking radioactive materials out of the plant. But Watras set off the radiation detectors on his way in! Not only that: the reactor had not yet been turned on, so the problem had to be something that Watras had brought in from his home, not from his job.

When the Philadelphia Power Company checked Watras's home in nearby Boyertown, it found levels of radon about 700 times greater than current federal safety standards. At the time, this level was by far the highest concentration of radon ever detected in a U.S. home.

Because the problem was clearly not related to the nuclear power plant, the Philadelphia Power Company quickly turned the problem over to the state. Although no one in the state's Environmental Resources Agency at first quite knew what to do, they responded quickly; within days, the Watras family had been evacuated from their home, not to return for over six months. Within a few weeks, state officials were testing local school buildings and offering free radon surveys to homeowners in the neighborhood.

The response to the survey offer was very low until the state started a newspaper and radio advertisement campaign in October 1985: in the first week, over 10,000 requests for surveys poured in, and roughly half the houses tested showed radon concentrations above the current recommended level. Very soon it became clear that this was not an isolated incident, but rather a large-scale problem involving a significant part of the population. Since then, both the media and the scientific world have started to pay attention to the dangers of radon. "Scientists believe that radon is responsible for as many as 20,000 lung cancer deaths each year," according to *The Washington Post*. It "may endanger 8 million homes," according to *The New York Times*. In a recent editorial, the *Times* talked about "a bizarre hazard of living indoors."

Yet before the Stanley Watras story broke in 1985, there was virtually no public awareness or interest in radon. This is graphically illustrated in Figure 1-3, which shows the number of stories on radon run by *The New York Times* each year since 1979. Until the end of 1984, interest was comparatively small; suddenly, in 1985, public interest shot up. Neither was there a

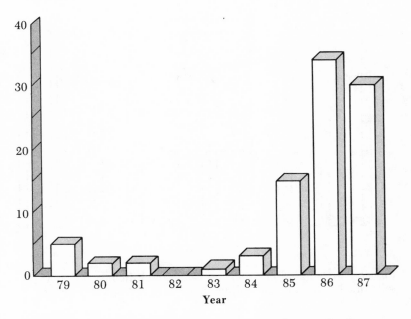

Figure 1-3 Number of stories each year on radon in *The New York Times*. Note the large increase around 1985.

great deal more interest on the part of scientists. Figure 1-4 shows how the number of scientific reports on the hazards of radon gas has followed a similar trend of a sharp increase starting around 1985.

What radon actually is has been reasonably well understood since the beginning of this century. In fact, even then, there was some concern about the hazards of radon. The very first nuclear physicist, Ernest Rutherford, wrote in 1907, "We must bear in mind that all of us are continuously inhaling the radium and thorium emanations and their products [i.e. radon]. . . . Some have considered that possibly the presence of radioactive matter and ionized air may play some part in physiological processes."

In fact, even after eighty years, much remains unknown about the hazards of radon gas. A recent *New York Times* headline neatly sums up the current situation: "Radon: Threat Is

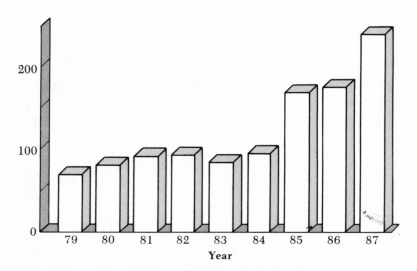

FIGURE 1-4   Number of papers on radon published each year in scientific journals. A marked increase around 1985 is noticeable.

Real but Scientists Argue over Its Severity." It is as well to put these uncertainties in perspective. A good deal is known about the hazards of radon: the disagreements are basically about exactly *how* hazardous it is. For example, the most recent estimates of the risks of radon were published by the National Academy of Sciences in 1988. Other recent estimates of this risk supplied by national and international agencies over the last ten years have ranged from three times smaller (1984) to twice as big (1980), with 1977 estimates about the same as 1988. The differences are significant but not enormous. So we have a rough idea of the size of the risk from radon, but we certainly do not know it exactly.

In this book we will look at the evidence on which these estimates are based. As we shall see, scientists disagree not only when they are looking at different types of evidence but even when they are considering the same evidence. So we must first establish, within limits, what the hazards of radon are.

The next question will be to find out if radon is a problem in *your* home. We will cover the best methods for doing actual measurements of radon in your house. Knowing the concentration of radon gas in your home, the next step becomes a matter of judgment: Do nothing? Take a few minor steps? Do something major? What, and at what cost?

But let us start at the beginning with radioactivity, the property of radon that causes all the trouble.

What is Radiation?

To understand radiation, we first need to know something about *atoms*, nature's building blocks. Literally everything is made out of these microscopically small pieces of matter measuring only a few billionths of an inch across. The human body, for example, consists of some billion-billion-billion atoms (this means 1 followed by twenty-seven zeros).

Suppose that we had a very powerful microscope and could see the fine details of individual atoms. The first thing we would notice, as in Figure 2-1, is that the atom is not solid inside like a billiard ball. It actually consists of even smaller particles, called *electrons*, moving around inside it, rather like the planets orbiting around the sun. Just as the sun is at the center of the solar system, at the center of the atom is the atomic *nucleus*. The nucleus is far smaller than the atom itself; for example, the nucleus of an oxygen atom is only about one

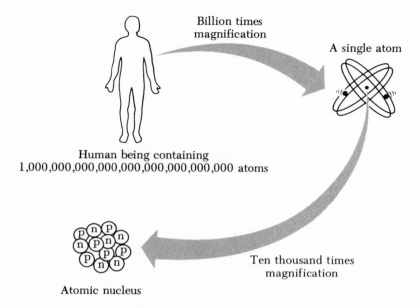

Billion times
magnification

A single atom

Human being containing
1,000,000,000,000,000,000,000,000,000 atoms

Ten thousand times
magnification

Atomic nucleus

FIGURE 2-1 The relative sizes of a human being, an atom, and the atomic nucleus.

ten-thousand-billionth of an inch across. But it is in the nucleus that the origins of radioactivity are to be found.

Inside the atomic nucleus are more particles moving about, not electrons this time, but two new types of particles called *protons* and *neutrons*. The number of protons in the nucleus is different for each chemical element. If an atom contains only one proton, it is an atom of hydrogen. If there are six protons, it is an atom of carbon; if there are eight, the element is oxygen. If there are eighty-six, it's radon.

Actually, this is an oversimplified view of nuclear structure. For example, neutrons and protons are held together in the nucleus by yet different particles (pions) and are themselves made up of still more particles (quarks). But describing the nucleus in terms of neutrons and protons is all we need to understand the basics of radioactivity.

To summarize

Everything is made of atoms.

Atoms consist of electrons and, at their center, the atomic nucleus.

The atomic nucleus consists of protons and neutrons.

The number of protons in the nucleus determines what type of chemical element the atom is.

Let us now turn to *radioactivity*. Some atomic nuclei are perfectly *stable*: consider, for example, the nucleus shown in Figure 2-2, which has eight protons and eight neutrons. As in Figure 2-2, we write it as oxygen-16 or ^{16}O. An important feature of this nucleus is that it will never spontaneously change its configuration. It will always look just the same way. In other words, it is stable.

Not all nuclei, however, are stable. Suppose we could look at an atomic nucleus of radon. We count the number of particles in it and find 86 protons and 136 neutrons. But if we keep looking at this nucleus, after a while something will happen, as illustrated in Figure 2-3: spontaneously, for reasons we will consider later, the nucleus has "kicked out" two protons and two neutrons stuck together in a tiny cluster. So what remains has 84 protons and 134 neutrons. Because the number of protons determines the chemical element — 86 was radon — what remains is a *different* chemical element. In this case it is polo-

The oxygen (O–16) nucleus:
- 8 protons
- 8 neutrons

FIGURE 2-2 Schematic view of the stable oxygen nucleus containing eight protons and eight neutrons.

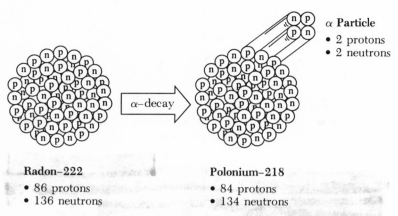

FIGURE 2-3 The alpha decay of radon-222, leading to the emission of an alpha particle. Note the radon has changed to another chemical element, polonium.

nium, discovered by Marie Curie and named after her native Poland. Quite spontaneously, a *nuclear reaction* has occurred. The tiny, tightly bound cluster of two protons and two neutrons coming out of the nucleus is called an *alpha particle* (after the first letter of the Greek alphabet, α).

Radioactivity, then, is a random, spontaneous change in the atomic nucleus that results in a small particle being ejected from it. The particles that fly out of the atom can sometimes hit sensitive parts of our bodies, producing cancer, a process discussed in detail in the next chapter.

The emission of α (alpha) particles is not the only type of radioactivity. Another type is the emission of *beta* particles (the next letter in the Greek alphabet). In beta decay (Figure 2-4), as it is called, one of the protons or neutrons in the nucleus spontaneously "converts." In other words, either a proton turns into a neutron, or a neutron into a proton. In either event, two particles are ejected from the nucleus. The first ejected particle is called a *neutrino*, a very strange object in that it seems to have no weight at all. The second ejected particle is an electron, otherwise known as a beta particle. This electron, unlike the ones shown in Figure 2-1, which orbit around the nucleus, is a newly created particle. It is created the moment the proton turns into a neutron or the neutron into a proton. In beta decay, the total number of neutrons and protons remaining in the nucleus stays the same: in the example in Figure 2-4, the total remains at 214.

There is a third type of radioactivity called *gamma* decay. Here a gamma ray—a type of x ray—is emitted from the atomic nucleus. Gamma rays and x rays are similar to ordinary light rays, except that they are invisible, more energetic, and more harmful.

In fact, the radioactive atoms that are of interest in this book emit all three types of radiation, but we shall be concerned mainly with alpha particles, which are much more hazardous than the other types of radiation.

The type of nucleus completely determines the kind, if any, of radiation emitted. For example, a nucleus with eight protons and eight neutrons (regular oxygen) will never radioactively decay; one with 86 protons and 136 neutrons (radon) will eventually undergo radioactive decay and eject an alpha particle (Figure 2-3); and one with 82 protons and 132 neutrons (lead) will eventually decay and emit a beta particle (Figure 2-4).

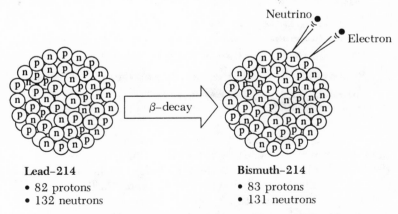

Neutrino

Electron

β–decay

Lead–214
- 82 protons
- 132 neutrons

Bismuth–214
- 83 protons
- 131 neutrons

FIGURE 2-4   The beta decay of lead-214, leading to the emission of a beta particle. Another ghostly particle called a neutrino is also emitted, but it is so unreactive that it would probably pass right through the earth without hitting anything.

Why do some types of nuclei radioactively decay while others do not? Think of the nucleus as a collection of balls (neutrons and protons) held together with "glue." The glue is the so-called nuclear force. Because only so much nuclear glue is available, if the nucleus contains too many "balls," there will not be enough glue to go around. Then the nucleus will not be stable and must get rid of some of its neutrons or protons. A lead nucleus consisting of 82 protons and 132 neutrons, for example, is a nucleus with insufficient glue to hold itself together. It is unstable and will emit particles. On the other hand, a lead nucleus with 82 protons and 126 neutrons has just enough glue to hold itself together; this is the stable form of lead that will not decay.

To summarize

Radioactivity is a spontaneous change in an atomic nucleus resulting in the emission of particles from the nucleus.

Alpha (α) decay is the emission of two protons and two neutrons from the nucleus.

In alpha decay, the two protons and two neutrons come out bound together in a tiny cluster called an alpha particle.

Beta decay is the emission of an electron from the nucleus. Gamma decay is the emission of an energetic x ray (or gamma ray) from the atomic nucleus.

Some atoms are stable; others spontaneously undergo different types of radioactive decay, depending on the type of nucleus.

When do radioactive nuclei spontaneously undergo alpha, beta, or gamma decay? It turns out that this question is beyond the bounds of present scientific knowledge. As far as we know, it is impossible to predict when a *particular* radioactive atom will decay. However, atoms come in extremely large numbers and we *can* predict, on average, how many of a large group of atoms will decay during any given time. To do this we use a property of the atom called its *half-life.* The half-life of a type of atom is the time it would take for half of a group of those atoms to undergo radioactive decay. Different types of atoms have different half-lives; the shorter the half-life, the faster the atoms will decay. Half-lives vary enormously from one type of atom to another — from billionths of a second to billions of years.

So there are two basic features that describe a particular type of atom's decay: the type of decay (alpha, beta, or gamma) and the speed of the decay (the half-life, usually written as $T_{\frac{1}{2}}$). To see how these ideas work in practice, let us look at the radioactive decay of radon-222, which contains 86 protons and 136 neutrons for a total of 222 particles. When radon-222 decays we have

$$T_{\frac{1}{2}} = 3.8 \text{ days}$$

Radon-222 $\xrightarrow{\hspace{3cm}}$ polonium-218 $+ \alpha$

What does this formula tell us? First, we see that radon-222 has undergone alpha (α) decay. Remember that an alpha particle is two protons and two neutrons bound together. As the radon-222 has lost four particles the product on the right has only 218 remaining. The radon has also changed in another way: it has lost two protons, so the new, smaller number of protons means that it has become another element, polonium.

Second, the half-life, $T_{\frac{1}{2}}$, specifies how fast the decays will take place. In our formula, the half-life is just under 4 days. If we start with, say, 1000 radon atoms, in 4 days half of them (500) will have changed by alpha decay into polonium, and 500 alpha particles will have been emitted. During the next 4 days, half of the remaining 500 atoms will decay, emitting another 250 alpha particles; then half of the 250 will decay. After every half-life, half of what remains decays away.

Before moving on to the biological effects of these particles we need to define one more concept: the *range* of an alpha particle. Like a car, which can go only so far before it runs out of gas, an alpha particle can travel only so far before it runs out of energy. The alpha particles that interest us in the context of radon have very short ranges: in air they can travel from about 2 to 3 inches before they stop, having lost all their energy. In human tissue, which is far denser than air, their range is much smaller still, between about 2 and 3 thousandths of an inch.

These limited ranges mean that to damage a living cell, an alpha particle must originate very close to that cell. The significance of this fact will become clear when we discuss radon in more detail in Chapter 4.

To summarize

The half life ($T_{\frac{1}{2}}$) tells us how long it takes for half of a group of a particular type of radioactive atom to decay.

The shorter the half-life, the more quickly the radioactive decays will take place.

Alpha particles have short ranges: a few inches in air, a few thousandths of an inch in human tissue.

An alpha particle must therefore originate very close to a cell if it is to damage that cell.

Why is Radiation Dangerous?

RADIATION causes cancer—that much is known for sure. As we shall see in Chapter 4, a good deal of evidence supports this assertion, including studies on survivors of the atomic bomb explosions at Hiroshima and Nagasaki.

Cancer is a major killer. In the United States alone, around one million new cancer cases are diagnosed every year. The most common type—and one of the most deadly—is lung cancer. A person's chances of being alive five years after diagnosis of lung cancer are currently less than one in ten.

To understand the nature of cancer, it is important to realize that a basic function of many cells in the body is to divide to allow for growth or replacement of different body parts. Cells with this capacity are called stem cells. Some examples are stem cells in the bone marrow, which continually divide to replace lost or damaged blood cells. Another example, which we will discuss in detail later, are basal cells in the lung that divide and replace cells in the lung airways.

A cancer occurs when something goes wrong with the division process. When a cancer, or tumor, is diagnosed, it usually appears as a large, loosely organized group of cells. These cells are not under the control of the normal regulatory mechanisms of the body. They are out of control and usually growing rapidly. Because of these regulatory mechanisms, normal cells "know" where they should be and what they should be doing. Cancerous, or malignant, cells are not under such control and will invade adjacent parts of the body, damage them, and prevent them from doing their job. Such damage often is the ultimate cause of death from cancer. In addition, many tumors will break up into pieces, called metastases, which will travel to distant parts of the body, start growing, and there cause damage and perhaps death.

The exact mechanisms by which radiation (or anything else, for that matter) causes cancer are not well understood. It is clear, however, that cancer is connected with the replication of cells, so the next part of the story is concerned with *cells*.

To summarize

Radiation causes cancer.

Cancer is a group of loosely organized cells that have had their growth-control mechanisms damaged. Typically, these cells multiply rapidly and invade neighboring parts of the body.

Lung cancer is the most common and one of the most lethal of all cancers, responsible for about 150,000 new cancer cases and 130,000 deaths each year in the United States.

Cancer is related to the division and replication of cells, the building blocks of all living things. A typical human cell measures about one-thousandth of an inch across—much bigger than any atom, but so small as to be invisible to the naked eye. At the center of the cell lies the cell *nucleus* (not to be confused with the *atomic* nucleus). Inside this cell nucleus is a tiny amount, less than one-trillionth of an ounce, of the chemical *DNA* (deoxyribonucleic acid). The DNA in the cell determines what kind of cell it is, where it should be, and what it should be doing — whether it is a liver cell, or a skin cell, and so on. And if the cell is, say, a hair cell, the DNA will also determine all the fine details — whether, for example, the hair is light or dark, or straight or curly.

The structure of DNA has been known since the 1950s. Figure 3-1 represents a strand of DNA magnified 10 million times. It consists of two spiral strands twisted around each other to form the famous "double helix." If we were to unwind the double helix, we would end up with something resembling a flat ladder — where the rungs and uprights are made of different organic chemicals. The four chemical "rungs" in DNA are arranged in different orders or codes, each carrying different instructions for the body's cells — a little like a complicated Morse code.

The DNA in cells has a second, related, task to perform. As we discussed before, all cells in the body are made by being

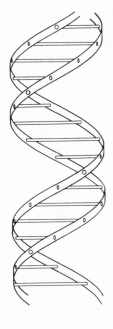

FIGURE 3-1   Schematic representation of DNA, the chemical containing all the genetic information for a cell.

reproduced from other cells. When a cell divides into two, all the information necessary for the "daughter" cells to know what *they* should be doing in the body is contained in the DNA. So the DNA must itself replicate to pass on this information to both the cell's daughters. Figure 3-2 shows this taking place: the two spiral strands are partially unwinding, and identical copies of the "parent" DNA are being made.

To summarize

All living matter consists of cells.

The nucleus of human cells contains the chemical DNA.

DNA determines the correct functioning and reproduction of the cell.

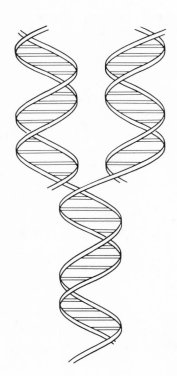

FIGURE 3-2   Schematic representation of the DNA self-replication process.

What can go wrong with this system if a human is exposed to radiation? Alpha (α) particles (and other types of radiation) can damage DNA. As an alpha particle passes through (or even near) DNA, it can cause a break in the strands of the DNA. At the bottom of Figure 3-3, you can see where an alpha particle has caused a break in a single strand of the DNA. Fortunately, cells are extraordinarily efficient at repairing these single-strand breaks. Chemicals called enzymes will quickly arrive at the scene and cut out the damaged piece. Then, by copying the other undamaged strand, they will make a replacement piece of DNA, put it in position, and glue the ends back together again, a remarkable achievement.

However, look at the damage caused by the alpha particle at the top of Figure 3-3. Here both strands, directly across from each other, have been broken. Now the repair chemicals that come along cannot look at the other strand to see what an

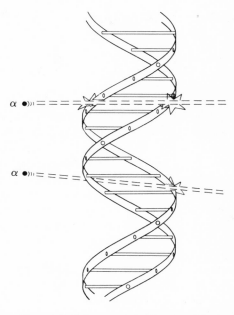

FIGURE 3-3 The effects of alpha particles hitting DNA. The alpha particle at the bottom has broken just one DNA strand, while the alpha particle at the top has broken both strands.

undamaged strand is like because that has been damaged as well. There is no "template" left to use as a model for making repairs. The enzymes, in this case, try to make repairs, but sometimes they will repair the damage incorrectly.

What are the consequences of such an incorrect repair? They can be catastrophic. If the DNA, which is responsible for the correct workings and reproduction of the cell, is permanently damaged, it may pass on this damage to the cell's "daughters." In other words, the cell and its offspring will contain *mutations*.

To summarize

Radiation is capable of breaking the strands of DNA.

When radiation causes double-strand breaks, the resulting damage may be permanent and may be transmitted to the cell's daughters. If this happens, the cell's daughters and their offspring are said to contain *mutations*.

Although mutations, by themselves, probably rarely cause cancer, they can be the first step on the road toward a cancerous cell. It is now thought that induction of cancer is a two-stage process, which is illustrated in Figure 3-4. At the top of the figure, a healthy cell (H) divides to produce two healthy daughter cells, which in turn divide normally. At the bottom of the figure, conversely, a healthy cell is exposed to radiation. In many situations, the radiation could so damage the cell that it could not divide at all. Such damage would not be too much of a problem because there would normally be plenty of undamaged cells remaining to do that cell's job. In the case shown here, however, the radiation produces a mutation (M) in the DNA of the cell, but without preventing the cell from dividing. Thus, the mutation is passed on to the cell's offspring. However, the cells are still under control and dividing normally. But suppose, later, that one of the cell's offspring is damaged again, say by some chemical such as tobacco smoke. Damage at this second stage can turn the already-damaged cell into a malignant cancer cell (C), in which all control of cell division is now lost.

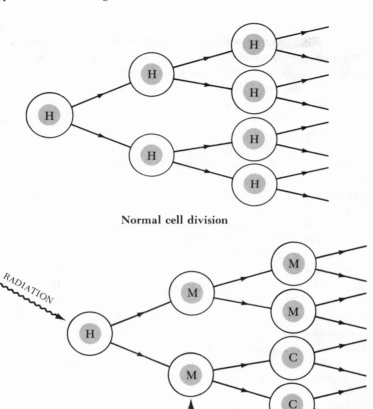

Normal cell division

RADIATION

CHEMICAL

| Initialization stage | Promotion stage | Cancer expression stage |

FIGURE 3-4 In normal cell division, a healthy cell (H) divides to produce two healthy cells, which in turn divide to produce more healthy cells. When radiation damages a cell, it can very occasionally produce a mutation (M), which is passed on to the cell's daughters. These mutant cells may then be in a "precancerous" state, and further damage, such as from a chemical, may convert such a precancerous cell into a cancerous cell (C).

These two different steps in cancer induction are usually called *initiation* and *promotion*. Radiation is typically an initiator: it can and does make mutations in the DNA of cells. But it is not a very efficient promoter. On the other hand, many chemicals, such as those found in cigarette smoke, are very good promoters. The link between the radon hazard and cigarette smoking — which we shall discuss at length later in the book — thus comes as no surprise.

This two-step model of cancer induction helps explain one of the main features of radiation-induced cancer: the so-called latency period. This latency period is the time between exposure to the initiator (radiation) and the development of cancer. For lung cancer, this can be a very long time, typically 10 to 25 years; this span is interpreted as the time that cells remain in a pre-cancerous state before being damaged a second time (for example, by cigarette smoke), which turns them cancerous.

Although this scheme is generally correct, some of the steps leading from DNA damage to cancer are only dimly understood. Very recently, new techniques in molecular biology have shed some light on the detailed mechanisms involved in initiation and promotion. In 1987, for example, researchers in Holland and England identified the particular part of the cell's DNA that gets damaged, causing lung cancer. They found that a tiny piece of a particular chromosome (the unit into which DNA is organized in the cell) was missing in most lung cancers. This missing piece probably contains the piece of DNA code that controls the cell's growth. Researchers around the world are now trying to identify what is in this missing piece of DNA. Although we do not understand *all* the steps leading to cancer, the link between radiation-damaged DNA and malignancy is now clear for most types of cancer.

To summarize

The formation or induction of cancer is a two-stage process.

In the first step, initiation, mutations are produced in the cell (for example, by radiation).

Mutated cells can stay in a controlled pre-cancerous state until damaged for a second time (promotion, for instance, by tobacco smoke), when they may turn cancerous.

So far we have been concerned with the effects of radiation on the individual actually exposed to that radiation. But there is another type of effect that we need to consider: genetic damage that we pass on to future generations. If radiation produces a mutation in the DNA of either a man's sperm cell or a woman's egg cell, the results would not show up in that individual, but they could appear either in the individual's children or in future generations. In other words, radiation damage to our reproductive or germ cells can be passed on to our offspring.

So far, no radiation-linked genetic effects have been observed in the children of people exposed to radiation, even at Hiroshima and Nagasaki, the sites of two massive atomic-bomb detonations in August, 1945. The reason that no such effects have been seen is basically because the natural frequency of inherited genetic mutations is so high: about one child in 20 is born with some genetic mutation. So a small increase in this high natural incidence would be very hard to detect. On the other hand, experiments with mice have clearly shown that the effects of radiation can be passed down from generation to generation.

Fortunately, as we shall see in the next chapter, alpha particles from radon and its daughters are very unlikely to have sufficient range to reach the germ cells in human reproductive organs. Henceforth, the effects we shall be concerned about are all to do with cancer in the individual exposed to the radon, and not in that individual's children.

To summarize

Radiation can produce genetic effects in the offspring of exposed individuals.

Alpha particles from radon and its daughters are unlikely to reach the sensitive germ cells of the body and thus cannot cause these genetic effects.

The important effect of radon and its daughters is lung cancer in the exposed individual.

Not all types of radiation are equally efficient at causing mutations and, ultimately, cancer. Many types of radiation (for example, alpha particles, beta particles, and gamma rays) can cause the types of DNA damage shown in Figure 3-3. However, alpha particles are the most effective at causing double-strand breaks, five or ten times more harmful than, say, x rays. Results such as these, however, come from laboratory experiments in a test tube. We cannot be sure that alpha particles also are more effective at producing human cancers. Still, it is reasonable (and prudent) to assume that they are.

Finally, we should note that not all cells in the body are equally sensitive to radiation. In 1906, two of the first radiation scientists, Bergonie and Tribondeau, found that dividing, or reproducing, cells were strongly affected by radiation, but that cells that did not divide were much less sensitive. This discovery is not surprising, for as we saw earlier, radiation damage is closely connected to cell replication and division. So body organs consisting mainly of cells that do not divide tend to be resistant to radiation. Examples are the brain, liver, kidneys, and muscle. Organs containing dividing cells, however, are very radiation-sensitive. Examples are bone marrow, skin, ovaries, and, as we shall see in the next chapter, the so-called basal cells lining the lung airways.

To summarize

Alpha particles are particularly efficient at causing double-strand breaks and mutations in cells. They probably also are very efficient at inducing cancer.

According to Bergonie and Tribondeau's law, cells that divide and replicate are likely to be much more radiation-sensitive than cells that do not divide.

What is Radon?
Where Does it
Come From?
Where Does it Go?

Some of the many forms of radon, ranging from radon-198 to radon-227, exist only in the laboratory for fractions of a second. (Recall from Chapter 2 that we use the number of neutrons and protons in the atomic nucleus to identify the form of an element.) Just one form of radon is primarily responsible for the health hazard: radon-222.

Radioactive atoms such as radon-222 decay to form different atoms and were, in turn, created by the radioactive decay of yet different atoms, in this case radium. Radium, in turn, was created by radioactive decay of still another atom, thorium. Figure 4-1 shows the whole chain of decays involving radon-222, which occurs about in the middle. The first atom in the chain is uranium-238, which is, essentially, a natural constituent of all rocks and soils. Uranium-238 occurs only in small quantities, usually a few parts per million. In other words, every pound of rock contains just a few millionths of a pound of uranium-238. As we shall see later, however, this amount is highly variable.

Uranium-238 is radioactive. Eventually, every atom of uranium-238 will decay to become a different atom. Figure 4-1 gives the two essential facts about this decay process: the type of decay and the half-life. The decay is by alpha-particle emission, from uranium-238 to thorium-234, with the four missing particles making up the emitted alpha particle. The half-life for this decay is more than 4 billion years, so the decays occur very infrequently, which is just as well for us. If the decays occurred more frequently, all the uranium in the earth would have decayed away by now and there would be no radon problem!

The alpha particles from the uranium-238 decays in rock and soil are not hazardous in themselves. As we saw earlier, the range of these alpha particles in solid materials is only a few thousandths of an inch. If the uranium is inside some rock and soil, the alpha particles created when it decays will certainly not have enough range to get out of the rock and into people. To see where the hazard comes from, we return to the decay chain in Figure 4-1. Once a uranium-238 atom decays to the next atom in the chain, the pace of the radioactive decays speeds up. Thorium-234 quickly decays by beta emission to protactinium-234, which in turn quickly decays to form uranium-234. Then three more decays occur, taking, on average, many thousands of years. Finally, radon-222 is formed.

Because all the atoms before radon-222 in the decay chain are solids, all the decays up to this point have taken place inside the rock, near the location of the original atom of uranium-238. None of the alpha particles released in the decays so far can get out of the rock. But radon-222 is different — it can escape because, unlike all the earlier atoms in the chain, it is a *gas*.

FIGURE 4-1   The radioactive decay chain starting from naturally occurring uranium-238 and involving radon-222. The letter next to the arrow indicates whether the radioactive decay is an alpha (α) decay or a beta (β) decay. The time under the name of the atom is its half-life. The two important daughters of radon-222, polonium-218 and polonium-214, are specially marked.

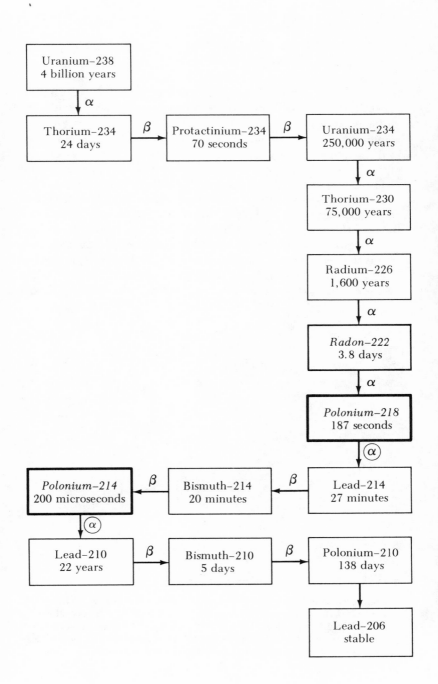

To see how this makes a difference, consider Figure 4-2, which shows a few grains of soil or rock—or perhaps pebbles—which contain some uranium-238. After a series of decays, a uranium-238 atom will be converted to a radium-226 atom, the precursor of radon-222. When the radium-226 decays, an alpha particle will come flying out in one direction and the remaining radon-222 will recoil in the opposite direction. This is similar to what happens when a gun is fired: the bullet (the alpha particle) goes in one direction, while the gun itself (the radon-222) recoils in the opposite direction. For the most part, radon-222 recoils and remains inside the solid grain of soil or rock. Occasionally, however, the radon-222 will recoil out beyond the edge of the grain or into a crack in the grain.

If radon were a solid, it would simply remain where it was, in the spaces between the grains of rock. Being a gas, however, it can permeate or diffuse through the spaces between the grains to reach the surface. Radon-222 has a half-life of almost

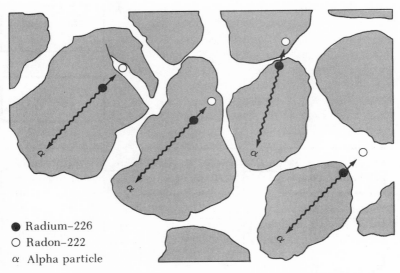

● Radium–226
○ Radon–222
α Alpha particle

Figure 4-2 Illustrating how radon-222 gets out of rock. When radium-226 decays, an alpha particle is emitted in one direction and radon-222 recoils in the opposite direction. Sometimes, as in the left-hand and right-hand decay, this recoil takes the radon-222 out of the rock and into the spaces within and around the rock.

4 days, which is long enough for it to travel a few yards through the ground. So as long as it was formed within a few yards of ground level, radon can escape into the outside air. Where the gas then goes depends on where the rock is. If it is outside, the radon-222 will mix with the open air; if it is under a house, radon can escape into that house.

To summarize

Uranium-238 is a natural, primordial component of rock and soil.

Uranium-238 is at the head of a chain of radioactive decays leading to radon-222 and beyond.

Because radon-222 is a gas, it can travel through the spaces between grains of soil to reach the surface.

Radon-222 is a colorless, odorless, almost totally non-reactive gas. It neither reacts with body tissue nor dissolves easily in body fluids. When it radioactively decays, it emits an alpha particle. Let us suppose you do inhale a radon-222 atom, which then floats into the lung. What will happen? Radon-222 has a half-life of 3.8 days, which is probably long enough for you to exhale it before it decays. On average, inhaled air stays in the lung less than a minute. The chances that a radon-222 atom will decay during this time are miniscule—about one in fifteen thousand. Even if the atom decays, the resulting alpha particle probably will not have enough range to reach the lung's sensitive parts.

Thus, radon-222 itself is not hazardous. However, when the radon gas decays by alpha-particle emission to produce polonium-218, that atom also decays by alpha-particle emission—with a half-life of about 3 minutes. But here is the difference: while radon-222 is a gas, polonium-218 is a solid. Almost as soon as they are formed, most of the polonium-218 atoms will attach themselves to tiny pieces of dust floating around in the air. No matter how clean or dirty the environment, air always contains plenty of these virtually invisible flecks of dust, called aerosols. When we breathe in air, aerosols are also swept into our lungs. As they enter the lung, these aerosols, some with polonium-218 attached, can get trapped on the surface of the lung, very close to radiation-sensitive lung cells. These trapped aerosols (and the attached polonium-218) stay in the lung for as long as 30 minutes — plenty of time for the polonium-218 to decay and emit an alpha particle. Because the decay occurs so close to the radiation-sensitive lung cells, the emitted alpha particle has enough range to reach and damage these cells.

The key difference, then, in the lung between radon-222 and its daughter, polonium-218, (both of which emit alpha particles) is that the gaseous radon-222 rapidly floats in and out of the lung before it has a chance to decay near the lung wall, but the solid polonium-218 gets trapped there. While the polonium-218 is attached to the wall of the lung, it can radioactively decay and emit its damaging alpha particle.

This lung hazard has yet another dimension. Returning to the decay scheme in Figure 4-1, note how polonium-218 quickly decays and turns into lead-214, then bismuth-214, and then polonium-214. This polonium-214 is another atom that decays by alpha-particle emission, resulting in dangers as serious as those from polonium-218. Again, when polonium-214 decays, it is likely to be attached to a dust particle sitting right next to the lung wall; therefore, the emitted alpha particle has a good chance of damaging the DNA of adjacent radiation-sensitive lung cells.

Taking a final look at the decay chain in Figure 4-1, there is one more alpha decay in the chain before it finally reaches lead-206, which is stable and does not decay. Before that final alpha decay, however, comes lead-210, with a half-life of 22 years. By the time it decays, it will almost certainly have been secreted from the body, so the final alpha decay in the chain is of little biological importance.

To summarize

Radon-222 is not itself hazardous because it does not spend enough time in the lung to decay and emit an alpha particle.

The radon-222 daughters, polonium-218 and polonium-214, attach themselves to dust particles (aerosols) that may be inhaled and settle on the surface of the lung.

The alpha particles emitted by polonium-218 and polonium-214 may damage radiation-sensitive cells near the surface of the lung, causing DNA damage, which can lead to cancer.

We have seen that radon-222 is the primary source of radioactivity in the lung through its daughters, polonium-218 and polonium-214. In fact, there is another natural decay chain that leads to *radon-220* (see Figure 4-3). This decay chain begins with thorium-232, which exists in rocks in concentrations similar to those of uranium-238.

Radon-220 is also a gas. It can travel out of rocks just like radon-222, and its daughters can be deposited in the lung. However, there is a difference. The half-life of radon-220 (see Figure 4-3) is less than one minute, compared with 4 days for radon-222. It therefore has a far shorter time, once it is formed, to get out of the ground. Not surprisingly, fewer radon-220 atoms get out of the ground than radon-222 atoms, and indoor concentrations of radon-220 are about five to fifteen times lower than those of radon-222. Consequently, the risk of lung cancer is probably much less than from radon-222 and we shall not consider it further. From now on, whenever we refer to "radon," we will mean radon-222.

To summarize

One natural decay chain goes from uranium-238 → radon-222 → radon-222 daughters.

A second decay chain goes from thorium-232 → radon-220 → radon-220 daughters.

Radon-220 atoms have a half-life of less than a minute, which is not enough time for them to escape from the ground in large quantities.

The danger from radon-220 is probably much lower than that from radon-222.

Which factors determine the levels of indoor radon gas and, consequently, the levels of radon daughters, polonium-218 and polonium-214, in our lungs? Basically, they are the following:

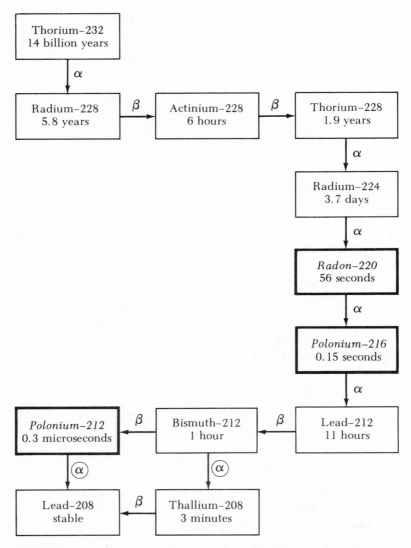

FIGURE 4-3   The radioactive decay chain starting from naturally occurring thorium-232 and involving radon-220. Two important daughters of radon-220, bismuth-212 and polonium-212, are specially marked. The biological significance of this decay chain is far smaller than the decay chain involving radon-222, shown in Figure 4-1.

- The rocks and soil on which we live.
- The houses in which we live.
- Our lungs, which breathe the air.

We shall look at each of these in turn.

ROCKS AND SOIL

Three basic factors determine how much radon will reach the surface and leak out of the ground.

- The uranium content of the rocks and soil below ground.
- How much radon can get into the spaces between the grains of rock when radium-226 decays (see Figure 4-2).
- How much radon can permeate up through the pores of the rock and soil to reach the surface.

As we have seen, radon-222, polonium-218, and polonium-214 are all "daughters" of the primordial uranium-238, which existed when rocks were first formed billions of years ago. Because a direct decay chain leads from uranium-238 to radon and its daughters, high concentrations of uranium in the ground will generally give rise to high concentrations of radon in the air above it. Conversely, low concentrations of uranium in the underlying rocks and soil will generally give rise to low concentrations of radon and its daughters.

Different types of rocks typically have different concentrations of uranium. Figure 4-4 gives some examples for the most common rocks and soil beneath houses in the United States and Canada. Although there are differences, different rocks contain, on average, quite similar amounts of uranium. However, the key word here is *average*. There may be very large variations from place to place and within a given type of rock.

For example, rocks that contain phosphates are often mined to obtain the phosphates for fertilizers. These rocks almost always have above-average uranium concentrations. For instance, two populated areas, each about 1,000 square miles, in north and west-central Florida, contain phosphate deposits

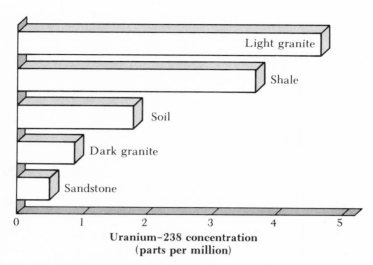

FIGURE 4-4 Typical concentrations of uranium-238 in various rocks and soils. (Adapted, with permission, from National Council on Radiation Protection and Measurements, Report No. 45, 1975.)

and high uranium concentrations. An even bigger area, about 135,000 square miles, containing high uranium levels is located mainly in southeastern Idaho and borders on Wyoming, Utah, and Nevada. Fortunately, this area is sparsely populated.

Occasionally, human enterprise is the cause of high uranium content in soil. An example is uranium mining. After extraction of the uranium from mined rock, the remaining waste material — known as tailings — was frequently used as an inexpensive base for building foundations. Although most of the uranium was removed during the extraction process, the other long half-life atoms in the decay chain (thorium-230 and radium-226) remain. They continue to decay, producing radon and its daughters. Well-known examples of this problem are in Durango and Grand Junction, Colorado, where the tailings were once used for the foundations of many homes and even schools. Other examples are in Monticello, and Salt Lake City, both in Utah.

Abandoned radium-processing factories are another man-made source of radioactivity because radium-226 decays to produce radon-222 and its daughters. Early in the century, radium was often painted on watch and clock faces because it causes phosphors to glow in the dark. When the demand for radium decreased these factories were abandoned, and the waste products were often used as land fill by builders. A well-known example of the problems caused by the use of abandoned radium-factory waste is in Essex County, New Jersey. There, more than seven hundred homes were built over radium-waste-contaminated ground.

On the whole, however, the man-made problem is tiny compared with the problem from naturally occurring sources because the man-made sources, by their nature, are small — a few square miles at most — whereas the natural sources may extend over far larger areas.

Returning to natural uranium-238, geologists have spent many years surveying the United States for uranium deposits for commercial mining. The results of these surveys are summarized in Figure 4-5. The shaded parts of this map show regions containing rocks with a high uranium content. One of the main sources for the information in Figure 4-5 is the National Airborne Radiometric Reconnaissance (NARR) program. In this study, started in 1974, aircraft criss-crossed the whole of the United States in lines about 3 to 6 miles apart. In all, over 1 million miles were flown. The planes (Figure 4-6) detected gamma rays from the radioactive decay of one of the daughters of radon-222, bismuth-214. When it decays, bismuth-214 emits gamma rays of a particular energy that travel far enough to reach and be recorded by detectors carried by the aircraft. The number of these gamma rays detected is then a measure of how much radon-222 or radium-226 is in the ground immediately below the aircraft. Some typical results of the NARR for eastern Pennsylvania and northern New Jersey are shown in Figure 4-7.

Until recently, there was sizeable debate about whether such maps are useful for predicting areas where houses are more likely to have high radon concentrations. It is now clear (see later in this chapter) that these geological maps are useful predictors of high-risk areas. Nevertheless, many homes with high levels of radon have been found in regions where high radon concentrations would not be expected, based on geologi-

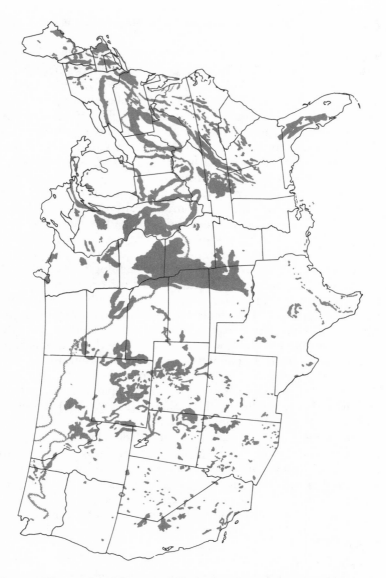

FIGURE 4-5 Areas with potentially high radon levels, based on geologic evidence. Houses in the shaded areas have a higher-than-average chance of having high radon levels. However, many houses outside these regions have high radon levels, and many houses in these regions do not have elevated levels. The shaded line shows the extent of glacial erosion. (Regions north of this area have very unpredictable radon levels.)

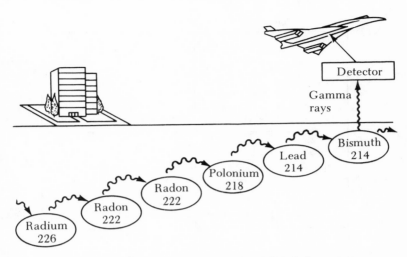

FIGURE 4-6 Illustrating the NARR technique for deducing the amount of radon-222 in the soil by measuring gamma rays (emitted by bismuth-214) with a detector mounted in an aircraft.

cal information. For example, one house with radon levels almost as high as in the Watras house (see Chapter 1) was found in what, according to the geological surveys, should be a low-risk part of Pennsylvania.

Even in high-risk areas, variations can be extremely local. For example, the nearest neighbors of the Watras family in Pennsylvania had indoor radon levels over a thousand times lower than those in the Watras house. When a local "hot-spot" shows up, such as the one at the Watras home, what appears to be happening is that the rock immediately beneath the house has fractured, allowing radon to move up very rapidly through cracks and to the surface. The extent of the cracks can be small enough to affect only one house.

One of the most notorious (and most studied) regions for high radon concentrations is the geologic region known as the "Reading Prong," which extends from Reading, Pennsylvania through New Jersey, New York, and up to Connecticut (see Figure 4-8). This region, which is mainly granite, has been known for many years to contain rocks with high concentra-

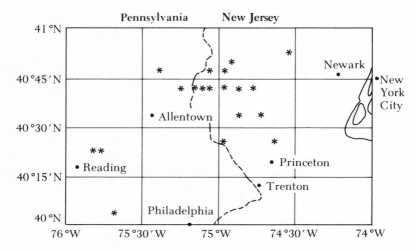

FIGURE 4-7 Results of the NARR survey of radium levels in the ground for northern New Jersey and western Pennsylvania. The aircraft covered this area with a criss-cross pattern with flight lines about 6 miles apart. Only areas with very high measured levels have been indicated with a star. (Adapted with permission from R.G. Sextro, et al., "Radon and Its Decay Products," Philip Hopke, Editor. Copyright © 1987, American Chemical Society.)

tions of uranium. Almost a quarter of a million homes are built on the Reading Prong, including the Watras house and their neighbor's house, which has a low radon level. Unfortunately, we have no way of making predictions of radon levels on a house-to-house basis.

To summarize

The amount of radon reaching the surface is determined by how much uranium is in the underlying rocks and by how efficiently the radon produced by this uranium can get to the surface.

Different rocks have different uranium contents, leading to different radon emission rates.

Some regions of the country have a natural abundance of rock with high uranium contents.

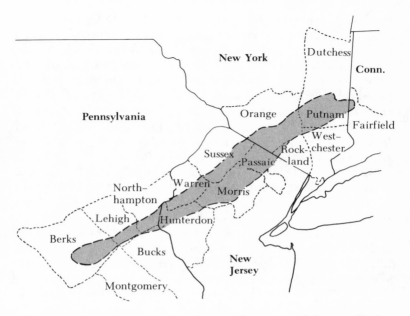

FIGURE 4-8   The shaded area indicates the extent of the so-called "Reading Prong." This geologic region, consisting mostly of granite, contains a much higher-than-average number of houses with high indoor radon levels. (Reprinted with permission from R.L. Fleischer, *Health Physics* 50. Copyright 1986, Pergamon Press plc.)

A smaller problem is radon emission from man-made sources, such as uranium-mining tailings or radium-painting waste products.

We now turn to the second factor determining how much radon will reach the surface. In Figure 4-2 we saw that radon-222 gets into the spaces or pores between grains of rock and soil by recoiling out of the solid grains. This recoil happens as radium-226 decays to radon-222 and an alpha particle. The alpha particle recoils one way, and the radon the other. Two factors affect how much radon can get into the spaces between the grains. One is the size of the grain: the recoiling radon atom has a tiny range, about two-millionths of an inch. If the grain is small, the radon has a better chance of starting out near the edge of the grain and getting out into the surrounding space. The second factor is water. Consider the decay of radium-226 to radon-222, shown on the top right of Figure 4-2. Here, the radon has recoiled right across a space between grains and has ended up in another grain. But if that space were filled with water rather than air, the radon would get stopped by the water before it could get across into the other grain. Water in soil therefore increases the amount of radon that gets into the spaces between the grains of rock or soil.

Let us consider the third factor that determines the amount of radon reaching the surface. We assume now that the radon has gone from the rock or soil into the spaces, or pores, within the rock. The next step is for the radon to permeate through these pores to reach the ground surface. Again, the main factor controlling this reaction is moisture: typically, radon atoms can diffuse about 6 feet up through the pores of dry soil. If the soil is moist, the radon might travel less than 1 foot. If the soil is very wet, the radon will probably travel no more than about an inch.

So moisture in the soil affects the amount of radon reaching the surface in two ways. First, increasing moisture content allows more radon to get out of rock and soil grains into the surrounding spaces. Second, increasing moisture content hinders radon from moving up through the pores to reach the surface. These two opposite effects are illustrated in Figure 4-9, which shows how much radon reaches the surface and escapes, in terms of the amount of moisture in the soil. As the moisture level increases, the amount of radon getting into the pores of the soil increases, so more reaches the surface. But when the soil gets even wetter, the radon cannot move through the soil and the amount reaching the surface decreases. For example, houses built on clay typically have low radon levels because, among other things, clays tend to have high moisture content. But, if the clay dries out, the amount of radon coming out of the ground will increase (see Figure 4-9). Moreover, if the clay dries out and cracks, then radon can simply permeate through the cracks.

With our present state of knowledge, we cannot examine underground soil and predict the amount of radon escaping

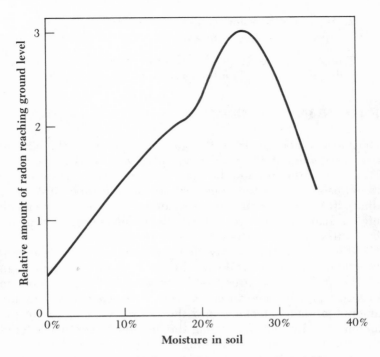

FIGURE 4-9   The amount of radon gas emerging from the ground changes as the soil gets wetter. When the soil dampens, the amount of radon reaching the surface goes up, but if the soil is very wet, much less radon reaches the ground. (Adapted with permission from E.P. Stranden, et al, from *Health Physics* 47. Copyright 1984, Pergamon Press plc.)

from the ground. The only way to know the level of radon in a house is to measure it.

To summarize

Moisture affects the amount of radon reaching the surface in two opposing ways:

1. High moisture levels increase the chance of radon reaching the spaces between grains of rock or soil

2. High moisture levels also prevent radon from moving up through these spaces.

There is no way to predict the radon levels in an individual house. The levels have to be measured.

RADON IN THE HOME

Outdoors, radon gas from the ground mixes with the air and gets very diluted because it has plenty of air to mix with. However, if the uranium-bearing rocks are under a house, the radon can leak into the house, where there is much less air to dilute it. So, in general, radon concentrations indoors are much higher than those outdoors. Typically, radon concentrations are five times larger indoors than out.

If a house had perfectly air-tight roofs and walls, the concentration of radon inside would build up to extremely high levels. In fact, all houses have some degree of ventilation, so eventually, a balance in concentration is reached where the amount of radon leaking into the house is balanced by the amount leaking out. *Increasing* the amount of ventilation (and thus the amount of radon leaking out) will *decrease* the concentration of radon in the house, whereas *decreasing* the amount of ventilation (making the house more airtight) will *increase* the concentration of radon.

There are various sources of radon gas in houses. We have emphasized radon coming from the ground, from rocks and soil. Let us look at some other possible sources.

- Building materials
- Outside air
- Tap water
- Natural gas

Figure 4-10 illustrates the relative importance of these other sources of radon in a typical home. They are all fairly small, in comparison with rocks and soil, for a "typical" house. Occasionally, however, any of them can be the cause of high indoor radon levels.

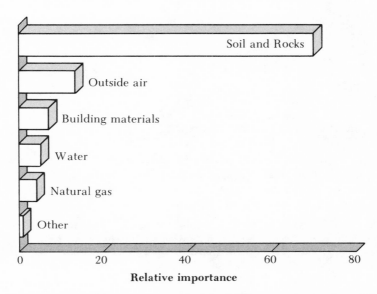

FIGURE 4-10   Radon enters a typical house from many sources. Of course a particular house will not necessarily exhibit exactly this pattern. (Adapted with permission from "Ionizing Radiation: Sources and Biological Effects," United Nations, 1982.)

Building Materials

Most walls and floors in buildings in the United States are made from either concrete, brick, gypsum, or wood. Figure 4-11 shows the relative speeds at which these materials typically emit radon. Concrete emits radon the fastest, wood the slowest. But there is no need to knock down your concrete walls and replace them with wood. From Figure 4-10, you can see that the contribution of building materials to indoor radon levels is usually about ten times less than that of the ground. Very occasionally, however, building materials do cause a problem. In Sweden, for example, until 1979, about-one third of all houses were built with concrete made from alum shale, a material that emits radon gas at about 40 times the rate of normal concrete. In this case, the building materials probably would be the largest source of radon in the home. No such high values have been reported in the United States.

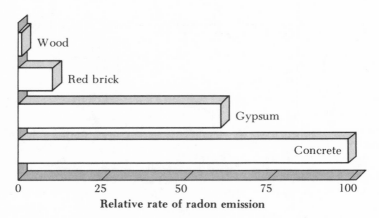

Relative rate of radon emission

FIGURE 4-11   Relative rates at which various building materials typically emit radon gas. (Derived with permission from J.G. Ingersoll, *Health Physics* 45. Copyright © 1983, Pergamon Press plc.)

Outside Air

Even without specific sources of radon in or around a house, radon will still come in from outside. As we have seen, radon concentrations outdoors are typically much lower than those indoors. Outdoor air is never a source of excessive amounts of radon in the home, and it usually has the effect of diluting and decreasing higher radon levels inside the home.

Tap Water

Normal tap water also contains radon. Actually, drinking tap water is not dangerous even if the radon level is quite high, because the alpha particles emitted by radon and its daughters in water do not have sufficient range to get out of the water. But there is another way that radon in water can be harmful: when water is used for everyday activities such as dishwashing, laundry, showering, and operating the toilet, radon will escape from the water into the air. The radon daughters can then be inhaled like any other source of radon. The higher the concentration of radon in the water, the more will escape into the air during these everyday activities.

Usually, the deeper underground the water source, the higher the radon concentration in the water. Water from surface reservoirs has very low radon levels; almost all public water sources have low radon levels, in part because the water is usually stored for some period of time before it is piped into houses. This storage time allows most of the radon to radioactively decay away. However, water from private wells can be a significant source of radon. Private well water usually comes from deeper underground and is not stored before being used.

About eight out of ten homes in the United States are served by reservoirs or public ground water. Of the remaining homes that get their water from private wells, about one in 20 have excessively high water radon levels. Consequently, about one to two million people in the United States may be affected by radon from water. We shall look into the radon-in-water problem in more detail in Chapter 10.

Natural Gas

The final source of radon in the home is natural gas. Radon dissolves in natural gas while the gas is still deep underground. When gas is burned indoors, the radon is released into the indoor air. Fortunately, natural gas is usually stored for some period between the time it is extracted from the ground and the time it is actually used. This gives the radon enough time to radioactively decay. As you can see in Figure 4-10, natural gas is a fairly insignificant source of radon in houses. No cases of excessive radon levels have been attributed to gas burning, so we will not consider it further.

To summarize

In the great majority of houses, the most important source of indoor radon is the ground under the house.

Occasionally, there can be a large contribution from private well water or, less commonly, from building materials.

Having established that rocks and soil are the most important source of indoor radon, we will take a closer look at the way radon actually gets from the ground into a house. First, we need to be able to talk more precisely about how much radon is in the air; we need to be familiar with the units in which radon concentrations are measured.

The idea is to take a volume of air and ask how many radioactive decays of radon-222 occur in it in say, one hour. The usual unit is called the picocurie per liter (normally abbreviated pCi/l). It corresponds to 133 radon-222 atoms decaying each hour in a liter of air, and producing 133 atoms of polonium-218. When we say that the radon concentration is 4 pCi/l, we mean that there are about 532 (i.e., 133 × 4) radioactive disintegrations of radon in every liter of air per hour. The number 133 is certainly curious and has its origins in the early part of this century, when radium was the standard for radiation measurements. In fact, authorities are trying to replace pCi/l with the new metric unit, becquerels per cubic meter; however, not many people are using the new unit.

So far we have been considering how to specify the concentration of radon itself. Because the real hazards are not from radon but from the radon daughters polonium-218 and polonium-214, we need to know the concentration of these two atoms in air. One atom of radon-222 decays to give one atom of polonium-218, which decays (through a couple of intermediate atoms) to give one atom of polonium-214. Although we might think that the concentration of each atom should be the same, several factors affect the concentrations of the daughters. One is ventilation: if the ventilation is strong, the radon will be flushed out of the room before it decays to produce its daughters. Another factor is dust size because, as we saw, polonium-218 and polonium-214 attach themselves to tiny pieces of dust, or aerosols. If these aerosols fall to the ground while a daughter atom is attached to them, their concentration in the air will be decreased. The extent of this decrease depends on the amount of dust in the room and the size of the dust particles, as well as the humidity.

The outcome is that the concentration of radon daughters is less than the concentration of radon itself. Consequently, there is a different unit for measuring the amount of radon daughters in the air. The unit that we use is called the "Working Level" (WL). This unit was originally a measure of radon-daughter exposure rate for uranium mine workers. (1 WL used to be the maximum level of radon daughters that was permissible in the air in a mine.) As we saw, there is no universal relation between the radon concentration (pCi/l) and the daughter concentration (WL); it depends on the conditions. For a typical house, 1 pCi/l of radon gas would give rise to about one two-hundredth of a WL of daughter products.

Our final unit arises because we would like to have a measure of how much exposure to radon a person receives over a period of time. Clearly, the longer a person spends in a radon-contaminated place, the greater the hazard. The unit is the "Working Level Month" (WLM). If a person were in a house containing 1 WL of radon daughters for 170 hours, he or she would be exposed to 1 WLM (170 hours is chosen because it is the number of hours that a miner normally spends underground each month). If the person spends twice as long in the house, he or she would be exposed to 2 WLMs. We will return to the WLM in the next chapter, when we relate radon-daughter exposure to the risk of cancer.

To summarize

Radon-222 concentrations in air are measured in picocuries per liter (pCi/l).

Exposure rates to radon daughters are measured in Working Levels (WL).

The amount of exposure over a period of time to radon daughters is measured in Working Level Months (WLM).

Having established units for measuring the comparing radon concentrations, we return to our original question, how does radon get into the home from the ground? The key is the force that drives radon-rich air out of the ground under the house and into the house itself. This force is the result of *pressure differences*. Whenever there is a difference in pressure between one region of air and another, there is always a flow of air from the high-pressure region to the low-pressure region. This flow is, for example, the cause of ordinary winds. It often happens that the air pressure in the house is slightly lower than that in the soil immediately below the house. The difference could be very small, perhaps less than one-hundredth of 1 percent. But this small pressure difference is enough to suck the radon-rich air out of the ground and into the house.

Two factors are responsible for this small but significant pressure difference. The first is wind. Wind blowing on the side of a house will create a tiny pressure difference between the air in the house and the air in the soil under the house. The pressure might be 76 centimeters of mercury directly under the house and 75.999 inside, but this small difference can be enough to pull the air from under the ground into the house. The second factor is temperature. Whenever a vertical wall separates two regions having different temperatures — such as an outside wall, where the inside is warmer than the outside — a small pressure difference is created. This effect is called the stack effect because it tends to make indoor air flow upwards, rather like hot air rising in a chimney stack. The net result is that air and radon are forced in at the bottom of the house.

These forces tend to drive radon-rich air into the house, but the amounts will vary with structural features. In houses with concrete floors, radon can enter through any crack or hole connecting the underneath soil to the interior of the house (for example, through cracks in a concrete slab or through badly sealed joints). Loose-fitting pipes are another common entry point, as is directly exposed soil, such as in a sump. Radon can also permeate through the floor itself, particularly a dirt floor, and even through the pores in concrete blocks. We shall talk more about these entry routes in Chapter 9.

To summarize

Radon is sucked up from the ground and into the house because of small pressure differences between air in the house and air under the house.

The primary causes of the pressure differences are wind and temperature.

Radon gets into the house through any crack or hole connecting the ground to the house.

Basements almost always have higher radon levels than rooms on other floors, partly because they are nearer the ground and partly because they tend to be poorly ventilated. For example, a recent study of New York and New Jersey homes found an average radon concentration of 2 pCi/l in basements, 1.1 pCi/l in first floors, and 0.9 pCi/l in second floors. Third and higher floors in apartment buildings all tend to have about the same low concentrations as second floors.

A group of houses that once attracted attention for having possible high radon levels were those with unpaved crawl spaces under the living area. Frequently, such crawl spaces were designed to be ventilated but were later sealed up as an energy-conservation measure. It was thought that such an action might cause a buildup of radon in the crawl space, which would then leak into the living area. However, more recent studies have shown that these houses are no more prone to high radon levels than houses built on a concrete slab.

This topic brings us to the question of ventilation and energy conservation. It has often been said that one of the main reasons for the current concern about radon is the energy crisis. The reasoning goes like this: as heating costs went up in the 1970s, people became more energy conscious and insulated their homes more efficiently. More insulation implies less ventilation and, so the reasoning goes, caused higher levels of radon. A 1984 headline in *The Wall Street Journal* summed up the issue: "Risk of Cancer from Radon Gas Increases with Growth of Energy-Efficient Homes." Let us see if the evidence supports this claim.

We measure the amount of ventilation in a house by how long it takes, on average, for all the air to leak out and be replaced by outside air. In a very energy-efficient home it might take about ten hours for one air change (one-tenth of an air change per hour). On the other hand, a very leaky house might have an air-change rate of two changes per hour. So the smaller the number of air changes per hour, the more tightly sealed the house. Several recent studies have tried to see if a connection really exists between energy-efficient houses and high radon concentrations.

Some typical results from these studies are shown in Figure 4-12. Here, 20 houses in Charleston, South Carolina, were surveyed both for their ventilation rates and their radon concentrations. Each point on the graph represents one house and shows its ventilation rate and radon concentration. If the ventilation rate really were an important factor, the points would show a trend: the fewer the number of air changes per hour,

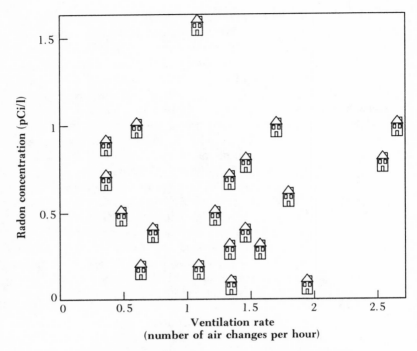

FIGURE 4-12   Results of simultaneous measurements of indoor
radon levels and ventilation rates in 20 houses in Charleston, SC.
If low ventilation rates (resulting from air-tight houses) implied
high radon levels, and vice versa, the points would be clustered
around a line going from the top left to the bottom right of the
graph. That the points are not clustered tells us that ventilation is
not a dominating factor in determining indoor radon levels.
(Adapted with permission from S.M. Doyle, et al, *Health Physics*
47. Copyright © 1984, Pergamon Press plc.)

the higher the radon concentration would be. The points would
be scattered about a line going from the top left (low ventilation, high radon) to the bottom right (high ventilation, low
radon). You can see, however, that no such trend exists in the
results in Figure 4-12. Similar results have been found in surveys in other U.S. cities and in Canada.

Ventilation does have some effect on radon levels. For
example, radon levels in U.S. houses in winter, when windows
tend to be closed, are typically about 40 percent higher than in

summer, when windows are open. Nevertheless, it is safe to conclude that ventilation is by no means a dominating factor in determining radon levels.

In fact, no single design feature of a house determines its radon level. Although a great many features can be important in certain circumstances, such features may be unimportant in other situations. A recent study looked at a host of possible factors in house design that might affect indoor radon levels. The study looked at the age of the house, number of floors, underfloor design, quality of the barrier between house and ground, draftiness, construction materials, ventilation, use of natural gas, and source of water. However, no strong connection was seen between high radon levels and any *single* factor. Sometimes one feature was important, sometimes another.

Of course, as we shall see later, if a home contains a high radon level, the house can certainly be altered to fix this. In this sense, house design and construction are important. But for houses not designed or altered with radon in mind, no particular feature is associated with high radon levels.

To summarize

In general, no single house-design feature is dominant in determining indoor radon levels; however, houses with high radon levels can be fixed with appropriate design modifications.

What mostly determines indoor radon concentrations is the nature of the rock and soil underground.

All of these considerations indicate that the only way to find out about radon levels in houses is to go out and take measurements in as many houses as possible. There have been two such scientific studies in the United States, one sponsored by the Department of Energy (DOE) and one by the Environmental Protection Agency (EPA). We will take a look at both of them.

The DOE study gathered information from over 800 houses in seventeen states. The *average* radon concentration in these homes was about 1.5 pCi/l. Not surprisingly, a wide variation occurred from house to house, a feature which is very significant. What we want to know is what fraction of houses in the

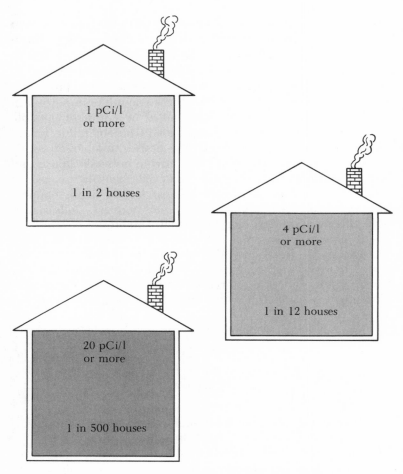

FIGURE 4-13   Some results of a study sponsored by the U.S. Department of Energy on indoor radon levels in U.S. houses. (Adapted with permission from A.V. Nero et al. *Science* 234, pp. 992–997, 1986).

United States have *high* concentrations of radon. The measured distribution of concentrations illustrated in Figure 4-13 shows what fraction of homes have *more than* a given concentration of radon. So, for example, about one in 500 houses have more than 20 pCi/l. One out of every 500 homes may seem a very small fraction. But because there are roughly 75 million homes in the United States, one in 500 means about 150,000 houses.

Because the DOE study reported on measurements from a rather small sample of houses, the Environmental Protection Agency (EPA) has embarked on a far larger study of twenty-five states and about 30,000 homes. As of the summer of 1988, this study was not yet finished (only seventeen states and 19,000 homes had been surveyed), but the preliminary results are quite revealing.

The results so far, broken down for each individual state, are summarized in Table 4-1. The last two rows show the results averaged over all 17 states, as well as the corresponding results from the DOE survey. It is interesting to compare the EPA (averaged over all seventeen states) and the DOE studies. The larger EPA study shows a much greater fraction of homes containing high levels of radon. The EPA study found that about 1 in 4 houses had more than 4 pCi/l, compared with about 1 in 12 as estimated in the smaller DOE study. This difference is a good illustration of the need for caution when looking at statistical surveys.

The EPA study has been criticized as being biased because the homes were not chosen in a completely random manner. The surveyed homes tended to be in known trouble areas where high radon levels might be expected. Consequently, the EPA is embarking on yet another study starting at the end of 1988. In the most definitive survey to date, 5,000 houses will be chosen completely at random and tested for radon.

The EPA survey also broke down the results by state into regions within each state. Figure 4-14 shows the region-by-region details of the results of the survey for 1,800 homes in Tennessee. From this survey, high indoor radon levels are most likely to occur in central and east Tennessee. But now look back at Figure 4-5. Here the shading shows regions with *potentially* high levels of radon, based on geological surveys of rock type. Sure enough, the predicted trouble areas in Tennessee are in the central and east regions. This link seems to hold true for almost all the states that were surveyed.

Several studies also have tried to look at the results of the National Airborne Radiometric Reconnaissance (NARR) program (see Figures 4-6 and 4-7) to see if the measurements

TABLE 4-1 *Results of the EPA survey of radon concentrations in houses in seventeen states*

	Number of homes surveyed	Average radon concentration (in pCi/l)	Highest radon concentration (in pCi/l)	Percent of houses with more than 4 pCi/l
Alabama	1200	1.8	180	6
Arizona	1507	1.6	51	7
Colorado	900	4.6	81	38
Connecticut	1500	2.9	81	19
Indiana	1217	3.6	72	26
Kansas	1000	2.9	27	21
Kentucky	900	2.8	66	17
Massachusetts	1659	3.4	61	24
Michigan	200	1.8	162	9
Minnesota	919	4.8	70	46
Missouri	1859	2.6	52	17
North Dakota	1596	7.0	184	63
Pennsylvania	429	6.2	114	37
Rhode Island	190	3.5	64	19
Tennessee	1800	2.7	100	16
Wisconsin	1200	3.4	142	27
Wyoming	800	3.6	81	26
All 17 states	18876	3.4	184	25
DOE study	817	1.5	26	9

0–10%   10–15%   15–20%   20–25%   More than 25%

FIGURE 4-14   Results of a study sponsored by the U.S. Environmental Protection Agency on indoor radon levels in Tennessee. The shading indicates regions with different percentages of houses containing more than 4 pCi/1. (Courtesy of the E.P.A.)

taken there are good guides to indoor radon levels in houses. The NARR program measured the level of bismuth-214 on or just below the ground, which should be an indicator of local outdoor (but not necessarily indoor) radon levels. There have been four studies (two in the United States, one in Canada, and one in Sweden) that have tried to see whether high readings from airborne measurements corresponded to high indoor radon levels near the same location. Three out of the four studies found a connection; one (in Pennsylvania) did not.

So it seems that geological evidence, such as that in Figures 4-5 and 4-7, is a good guide to *regions* that are at high risk. But, as we have seen, such evidence does not *guarantee* that low indoor radon levels will occur in individual houses in "low-risk" areas or, conversely, that high levels will occur in individual houses in "high-risk" areas.

To summarize

Based on surveys, the average indoor radon concentration in U.S. houses is probably a little over 1 pCi/1.

About one in 12 homes surveyed had indoor levels above 4 pCi/1. If this were true nationwide, about 7 million homes would have such levels.

About one in 500 homes surveyed had radon levels above 20 pCi/l. If this were true nationwide, about 150,000 homes would have such levels.

Geological maps (such as Figures 4-5 and 4-7) are available that predict which geographic regions will have above-average numbers of houses with high radon levels. Surveys of houses, by and large, confirm these predictions.

RADON AND THE LUNGS

A simplified diagram of the lungs is shown in Figure 4-15. Lungs are primarily concerned with breathing. Their basic function is to extract fresh oxygen from inhaled air and then to pass used carbon dioxide to the exhaled air. Air is breathed in through the mouth and nose and flows down the throat into the windpipe or trachea, shown at the top of Figure 4-15. From there it flows down a series of branching airways known as the bronchial tree. At the end of the branches of the tree, oxygen flows into the blood supply and waste carbon dioxide is removed from the blood.

Each of us inhales about 5,000 gallons of air each day. This air contains a host of pollutants, from dirt and dust to chemicals and micro-organisms. In order to keep the lungs clean and sterile, the body uses several defense mechanisms to filter out and then remove the dirt. How efficient these defense mechanisms are can be seen from the fact that average people inhale around a pound of dust and dirt over their lifetime, almost all of which is filtered and removed from the body.

The first line of defense is the nose, which filters out relatively large pieces of dirt and dust. These are then removed by sneezing and nose blowing. However, the aerosols that carry radon daughters are small and easily slip by the nose. The next line of defense is in the trachea and the bronchial tree. Figure 4-16 shows the inside surface of the lining of the bronchial tree. What actually traps the dirt is a sticky layer of mucus on the innermost surface, not visible in the pictures. Particles traveling through, such as aerosols carrying radon daughters, get stuck to this sticky layer.

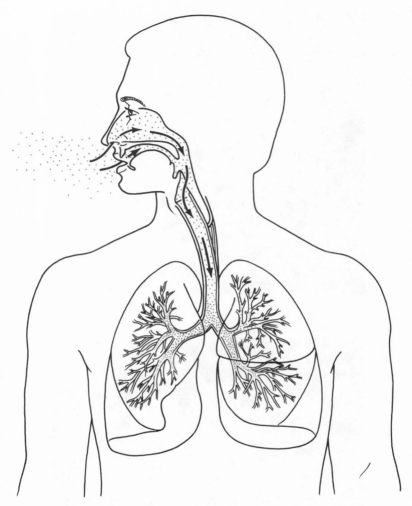

FIGURE 4-15 Schematic illustration of the deposition of radon daughters in the lung.

Below this mucus are tiny hairlike objects, called cilia, shown in Figure 4-16. These cilia bend forward and backward in waves, like corn in a windy field, to move the mucus up and out of the lung. Any particles stuck to the mucus will be removed at the same time. It takes between 3 and 30 minutes

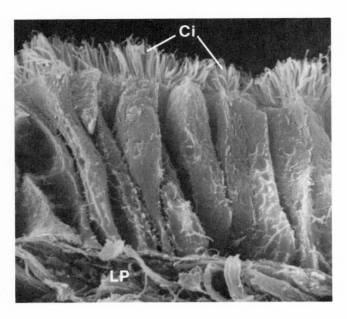

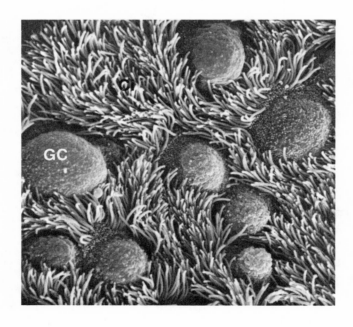

for mucus to move out of the lung, depending on its depth in the lung. Once the mucus gets out of the lung and into the back of the throat, it is either coughed out or swallowed.

Figure 4-16 shows larger cells, called "goblet" cells, below the cilia. Their job is to make mucus and secrete it to the surface. In addition to goblet cells, the surface of the lung contains ciliated cells, forming the base of the cilia. Neither type of cell divides or reproduces. These cells are continually flaking off and being carried away by the mucus. The job of replacing them belongs to a third type of cell in Figure 4-16, the *basal* cells. "Basal" simply means bottom; these cells lie beneath the goblet and ciliated cells and are also the base cells from which the others grow. Basal cells are continually dividing to replace goblet or ciliated cells that are lost.

Recall the law of Bergonie and Tribondeau, which we discussed in Chapter 3: dividing cells are much more sensitive to radiation than cells which do not divide. So it is quite likely that the basal cells are the crucial ones — the ones that may ultimately lead to lung cancer if they are damaged by radiation. We shall have more to say about this later.

Thus far we have a picture of radon daughters attached to dust particles or aerosols that are inhaled and come floating into the bronchial tree. The aerosols get trapped on the mucus layer on the inner surface of these airways. The mucus (with the attached particles) then gets moved out by the beating cilia. However, while this clearing out is taking place (taking up to 30 minutes), the radon daughters have a chance to radioactively decay. Recall from Figure 4-1 that the radon daughters have short half-lives, so there is a good chance that they *will* decay before they are moved out of the lung.

If they do decay, the alpha particles from polonium-218

FIGURE 4-16   Two photographs, enlarged about 20,000 times, of the wall of the bronchial tree in the lung. The photographs show the cilia (Ci), which beat back and forth to move mucus (not shown) out of the lung. Below them are various cells, including goblet cells (GC), which manufacture mucus, and basal cells, which are just above the bottom of the lung wall (marked LP).
(Reproduced with permission from *Tissues and Organs: A Text-Atlas of Scanning Electron Microscopy* by R.G. Kessel and R.H. Kardon. Copyright © 1979, W.H. Freeman and Company.)

and polonium-214 have ranges through human cells of 1.9 and 2.8 thousandths of an inch, respectively. The basal cells are, on average, about 2.2 thousandths of an inch below the mucus, where the daughters are trapped. This means that a good number of the basal cells are within range of the alpha particles emitted by the radon daughters and may be damaged by them.

To summarize

Inhaled aerosols carry radon daughters into the lungs, where they are trapped by the sticky mucus on the surface of the bronchial airways.

While they are attached to the surface of the lung, the radon daughters have a high probability of decaying and emitting an alpha particle.

The sensitive target cells for lung cancer are probably basal cells, located just beneath the inside surface of the bronchial airways.

Many of the basal cells are within the range of the alpha particles emitted by radon daughters.

While many factors determine how many radon daughters actually decay close to the radiation-sensitive basal cells, the only factor we can fully control is the concentration of radon in the indoor air.

For a given a concentration of radon daughters in the air, many factors determine how many alpha particles actually reach the sensitive basal cells in the bronchial tree. Among them are the person's age and breathing patterns, the size of the aerosol particles, the speed with which the mucus clears the aerosols out of the lung, and the depth of the basal cells below the mucus layer. As we shall see later, most of our As we shall see later, most of our estimates of the dangers of radon are based on workers exposed to radon in uranium mines. If any of the factors determining how much radiation from the air reaches the lung are different in the home than in a mine, risk estimates derived from miners may not be applicable to people in houses. Let us look briefly at the most important factors.

The Aerosol-Size Distribution

Most radon-daughter atoms get attached to aerosols or dust particles as soon as they are formed. One crucial factor in determining the amount of radiation reaching the lung is the size of these tiny aerosols. The smaller the aerosol particles, the faster they travel, and the faster they travel, the more likely they are to be deposited in the lung. In general, a house contains a variety of different-sized aerosols. A range of sizes, measured in a typical house, is shown in Figure 4-17. An average-sized aerosol is about four millionths of an inch across. Aerosols come from a variety of sources, such as smoking and cooking. Smoking actually tends to produce larger aerosols and cooking smaller ones. In mines, aerosol sizes tend to be about twice as large as those in the home.

Basically, the smaller the aerosol particle, the bigger the chance it has of being deposited in the lung. For typical aerosols in houses, about 10 percent of all aerosols that are inhaled will end up on the lung. In mines, where aerosol sizes are bigger, the fraction of inhaled aerosols which are deposited on the lung is smaller, around 3 percent.

Smaller aerosols are more likely to get trapped in the lung, which brings us to the second factor affecting the radiation dose to the lung: the *unattached fraction*.

The Unattached Fraction

You can see from Figure 4-17 that while most of the particles are around four millionths of an inch in size, some are far smaller. These smaller particles make up the so-called unattached fraction. They are actually atoms of the radon daughters that have not attached themselves to dust particles. As we saw, the smaller the particle, the more likely it is to be deposited in the lung; these "unattached" small particles therefore can contribute significantly to the radiation dose to the lung. As you might expect, one of the main factors determining how many daughters do or do not attach themselves to aerosols is the number of the aerosols in the air. The dustier the air, the more aerosol particles there are to which the radon daughters can attach. For example, in a dusty mine there are relatively few unattached atoms, whereas in a house there might be as many as one unattached atom for every twenty attached.

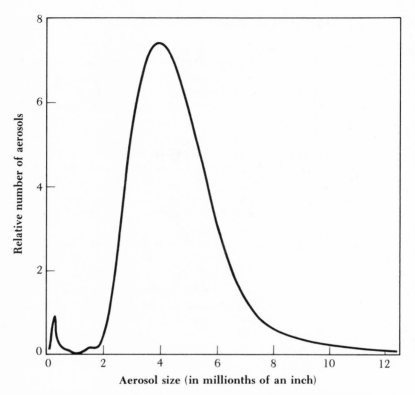

FIGURE 4-17   Relative numbers of different-sized aerosols present in a typical house in New York. Most of the particles are around 4 millionths of an inch, but there is a significant number of much smaller aerosols. (Adapted with permission from A.G. George and A.J. Breslin, "The Natural Radiation Environment III," T.F. Gesell and W.M. Lowder, Editors, Department of Energy, 1980.)

The Effects of Age

The age of the individual also affects the amount of radiation from the outside air reaching the lungs, mainly because the shape of a child's lung is very different from that of an adult. In addition, children's breathing patterns are different from those of adults; the crying of a newborn or the attempts of a baby to crawl are hard work! Figure 4-18 shows the relative doses to

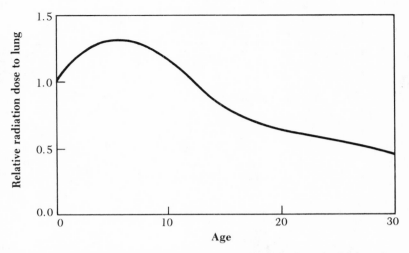

FIGURE 4-18 The relative radiation dose deposited in the lungs of people of different ages breathing in the same amount of radon daughters. The different shape of children's lungs results in children receiving a higher radiation dose than do adults. (Adapted with permission from W. Hofmann, *Health Physics* 37. Copyright © 1979, Pergamon Press plc.)

the lung (for the same amount of radon in the air) for children and adults. For example, the radiation dose to a five-year-old child's bronchial tree is almost three times the dose to an adult breathing in the same amount of radon.

Location of Sensitive Cells

The location of the sensitive cells is probably the least well understood of all the factors affecting the amount of radiation getting from the outside air to the sensitive cells in the lung. As we saw, the basal cells may be the sensitive "target" cells for lung cancer. They are the ones that divide to produce the other cells in the lung wall. On average, they are a little over two-thousandths of an inch below the surface of the lung wall. In fact, different people have their basal cells at different depths below the surface of the lung wall.

This range of depths is illustrated in Figure 4-19, which also shows the ranges of the alpha particles from the decays of

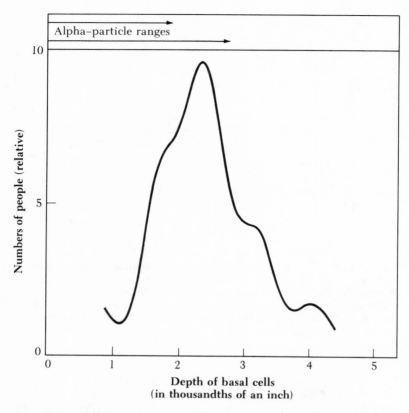

FIGURE 4-19 Relative numbers of people whose radiation-sensitive basal cells are at various depths below the surface of the lung. The average basal-cell depth is around 2 to 3 thousandths of an inch, but in some individuals the basal cells are 4 or more thousandths of an inch below the surface. The range of alpha particles from radon daughters in the lung is just under 2 or 3 thousandths of an inch, as shown at the top of the figure. Consequently, alpha particles can usually reach the sensitive basal cells.

the two radon daughters. If an individual's basal cells are quite deep below the lung surface, he or she might be shielded from all radon effects because the alpha particles, starting at the surface of the lung, simply would not have enough range to reach the sensitive basal cells. Someone with average-depth basal cells might be shielded from one but not the other of the radon-daughter alpha particles, whereas someone with shallow basal cells might be affected by both alpha particles.

It is not certain that the basal cells are the only sensitive "target" cells involved in lung cancer. Some evidence suggests that there are other cells in the wall of the lung that, if damaged, can initiate cancer. If this is true, then the depth of the basal cells is not important at all. This crucial point is still being investigated.

To summarize

The factors that affect the proportion of radon daughters in the air that reach sensitive parts of the lung include

1. the sizes of the aerosol particles
2. the fraction of radon daughters that do not attach to aerosols
3. the age of the person
4. the location and depth of the sensitive "target" cells beneath the surface of the lung wall.

We have looked at the main factors that control the proportion of radon daughters from the outside air that reach the sensitive parts of the lung. Unfortunately, we have little control over any of these factors except the obvious one, the concentration of radon daughters in the air. When we come to consider how to reduce the radon hazard, we will be looking primarily at the reduction of indoor radon levels. The exception is the issue of smoking, which we shall consider at some length in a later chapter.

But first, in the next chapter, we turn to the rather strange history of radon.

The Checkered History of Radon

THE effects of radon have been an issue for at least four hundred years. Over the years, radon has been considered both beneficial and harmful. In this chapter we will take a brief look at radon as it has slipped in and out of favor over the years.

The first known report of the hazards of radon was issued in 1556. It came in *De Re Metallica*, Georgius Agricola's classic text on mining. One section of the book is devoted to the health effects suffered by miners, particularly in the silver mines of the Erz mountains on the border of Germany and Czechoslovakia. The miners, Agricola said, breathed in a corrosive dust that "eats away the lungs . . . hence in the mines of the Carpathian mountains women are found who have married seven husbands, all of whom this terrible consumption has carried off to a premature death." Agricola also writes about another hazard of mining, namely "demons of ferocious aspect" who are "expelled and put to flight by prayer and fasting"—Agricola may not have been a completely reliable source! The lung disease described by Agricola, which miners call "Bergkrankheit," or mountain disease, was later shown to be lung cancer. As we shall see, the Erz mountain mines contained very high levels of radon, so it is likely that many of the "seven husbands" died of radon-induced lung cancer.

Radiation itself was not discovered till much later. Roentgen first discovered x rays in November 1895; Becquerel discovered radioactivity (in uranium) shortly after, in February 1896. Alpha radiation was found two years later by Ernest Rutherford, who examined the radiations coming out of uranium and identified both alpha and beta rays.

It did not take very long for the harmful effects of radiation to be observed. Thomas Edison, in March 1896, reported eye irritation from experiments with x rays—although, in fact, it was probably just simple eye strain. Ironically, because of this largely imagined eye problem, Edison stopped working with x rays, but he used what he had learned from his experiments to develop the first fluorescent light. By the end of 1896, however, harmful effects of x rays had definitely been observed, particularly skin burns and hair loss.

At this point, Marie Curie entered the story. She and her husband, Pierre (see Figure 5-1) were examining the uranium mineral pitchblende, the same material mined in the Erz mountains and discussed by Agricola over 300 years earlier. In Agricola's time the mines had been used to obtain silver, but by the 1890s, uranium was being mined as a pigment to tint glass.

The Curies found that the level of radioactivity in pitchblende was far too high to be caused by uranium alone. They concluded that two more radioactive elements had to be present. In this way they discovered first polonium, which they named after Poland, Marie Curie's homeland, and then radium.

Marie Curie isolated radium from the ore pitchblende. She bought about eleven *tons* of waste pitchblende from the Erz mountain mines and, after four years of work, she had isolated from it about three thousandths of an ounce of radium. Tragically, both Madame Curie and her daughter Irene would later die of radiation-induced cancer.

FIGURE 5-1   Marie Curie at work with her husband, Pierre. They were responsible for discovering polonium and then radium. Madame Curie ultimately died of leukemia, probably caused by radiation. (Photograph kindly provided by the Niels Bohr Library of the American Institute of Physics, New York.)

To summarize

1556   Agricola reports lung cancer in miners
               exposed to radon gas.

1895 – 1898   X rays, radioactivity, and alpha, beta, and
               gamma rays are discovered.

1898   The Curies discover two new radioactive
               elements, radium and polonium.

1899   Adverse effects of radiation are well
               established, especially skin burns.

When radium-226 decays it becomes radon-222 (see Figure 4-1). Radon was discovered in 1900 by Ernest Dorn, a German chemist, who found that radium gave off an "emanation" which, not surprisingly, he called "radium emanation." A year later, Rutherford showed that the "emanation" was, in modern language, radium decaying to give radioactive radon gas. This gas was chemically isolated in 1908 and called niton, from the Latin *nitere* (meaning "to shine"). Neither name stuck, however, and since around 1923, the name "radon" has been used.

As we saw, scientists recognized early on that radiation could be harmful and could damage human cells. Indeed, by 1899, the first malpractice award for x-ray burns had been made. So it seemed a logical idea that drinking dissolved radium might help *destroy* stomach cancer. In fact, drinking radium and radon-containing water dates back to Roman times. Radon spas had been popular ever since then, but now that a "scientific" explanation for their supposed efficacy was at hand, they became tremendously popular. In these spas, the radon was either directly inhaled in "ematoria" or dissolved in water and drunk. Various devices, called emanators, were sold to increase the concentration of dissolved radon in water, creating an artificial spa. A typical emanator is shown in Figure 5-2. Other emanators had suitably healthy sounding names such as "Vitalizer," "Revigator," and "Medicinal Strength."

After about 1903, radon water became more and more popular (see Figure 5-3), and this fad was fueled by the support of most doctors at the time. One wrote in a medical journal in

FIGURE 5-2 The "Radium Vita Emanator," which was on sale to the public in the 1930s. It was a device for dissolving radon in water, which was then supposed to be therapeutic when drunk. (Reproduced with permission from A.M. Jelliffe and F. Stewart, British Medical Journal, 1969.)

1916 that it "has absolutely no toxic effects, it being accepted as harmoniously by the human system as sunlight by the plant." In the same year, the American Medical Association refused to endorse several emanators because they produced too little radon!

Ailments that were suggested as curable by the radon waters included old age, insanity, bad breath, blindness, tuber-

FIGURE 5-3   Advertisement for "Agua Radium"—radium dissolved in water, which was sold to the public during the 1920s and 1930s. The instructions for this particular brand suggest mixing it with milk to make it "more digestive for all children and invalids." (Reproduced with permission from A.M. Jelliffe and F. Stewart, British Medical Journal, 1969.)

culosis, and malnutrition. A promotional catalog for one of the emanators claimed that radon "revives, regenerates and rejuvenates the entire human organism." A newspaper even speculated that radium might raise the dead, an idea immortalized in the 1935 movie serial "The Phantom Empire," in which Gene Autry was brought back to life in a "radium reviving room."

As well as breathing radon in ematoria and drinking it in water, the public ingested it in many other ways, especially in patent medicines. Happily, many of these, such as "Radiol," marketed in the 1900s, contained no radioactive material at all. Many products did contain radium (and therefore also radon), ranging from chocolate candies and bread to slippers to a toothpaste called "Radiogen," which supposedly would release a constant amount of radon as the teeth were brushed. When a German chemical company started stockpiling radioactive materials during World War II, Allied scientists became very concerned until they discovered that the company was preparing to manufacture radioactive toothpaste after the war!

The most notorious purveyor of patent medicines that contained radium and radon was William Bailey. Bailey was first heard of in New York City in 1915, selling automobiles. A few years later he was making and selling "Radiothor," a liquid tonic containing extremely high concentrations of radium. One of Bailey's most enthusiastic customers was Eben Byers, a well-known, wealthy Pittsburgh industrialist. From 1926 to 1931, Byers drank a bottle of Radiothor every day, totalling about 1400 bottles. By 1931, he was suffering from cancer of the jaw, and in April 1932 he died in great pain.

Byers was by no means the first to die from the intake of radioactive materials. In 1904, Pierre Curie had performed experiments with animals that breathed in high levels of radon gas; all of them rapidly died. In 1912, a German woman who had been injected with radium for "medical" reasons (to help her arthritis) died a month after her treatment. But, more than any other, Byers' death caught the public's attention and there was a great outcry.

Almost at the same time, another tragedy was unfolding. From about 1917 on, in New Jersey, around 2,000 young girls were employed to paint luminous dials onto watches. As they painted, the girls would lick their brushes to get a sharp point, swallowing the radium in the paint. By 1924, local dentists were seeing an epidemic of "jaw rot," and by 1929, it was clear that radium was the cause. Most of the girls subsequently died of anemia or bone cancer.

The heyday of the radium water emanator was in the mid-1920s. About 150,000 devices were sold in 1926. In the early 1930s, largely because of adverse publicity from the cases of Eben Byers and the radium-dial workers, the use of radium water started to decrease. As late as 1953, however, a Denver company was still advertising a contraceptive jelly that contained radium.

Radon spas, on the other hand, are still popular in many places. In Montana, for example, many disused mines that contain high levels of radon gas are open to the public. Visitors (see Figure 5-4) sit in the mines—which have exotic names like Merry Widow or Free Enterprise—or drink the water, or simply run the water over afflicted areas. In the Soviet Union, radon bath therapy is even more popular and is available through the National Health Service; about 25,000 baths per day were prescribed in 1972. The Pyatigorsk Superior Radon Center in Georgia, a resort region in the Soviet Union, treats more than 1,000 patients every day. There the radon is administered in a variety of ways, including drinks, baths, radon water enemas, and vaginal and nasal irrigations. Recently, a large new radon bath facility was installed in Moscow.

The risks of radon are known, but we have almost no evidence that radon baths or spas have any therapeutic value; virtually all of the studies made show no benefit. One exception is a 1979 Austrian study that suggested that low levels of radon stimulate the chemical enzymes responsible for DNA damage repair in certain cells (see Chapter 3). Even if radon does stimulate such repair, the hazards of radon far outweigh any conceivable benefits.

To summarize

Radon was discovered as "radium emanation" in 1900.

From 1900 to about 1935, the medicinal use of radium and radon was widespread and encouraged by the medical profession.

FIGURE 5-4   Visitors sitting in a mine in Montana that has a high radon concentration. It has been claimed that sitting in such mines, or drinking the water, brings relief for a variety of illnesses, such as arthritis. There is, however, little evidence to confirm this. (Photograph courtesy of Eugene Fischer.)

Little scientific evidence for any benefits from the intake of radium or radon has ever been presented.

The deaths of Eben Byers and many radium-watch-dial painters were the turning point in demonstrating the adverse effects of radium and radon ingestion.

Today, in radiotherapy, the cell-killing properties of radon daughters have been widely and successfully used for treating cancer. The idea is to enclose the radon gas in a small "seed," which is then implanted in the cancerous tumor. This procedure maximizes the radiation damage to the tumor and mini-mizes damage to the surrounding healthy tissues. The treat-ment is in contrast to conventional external radiotherapy, in which the radiation must pass through skin or healthy tissue on its way to the tumor. External treatment is often limited by how much radiation the healthy tissue can tolerate. Radon implants avoid these problems.

The first use of locally implanted radon seeds was reported in 1914 on a woman with inoperable jaw cancer. During the 1930s, more than 100,000 patients were treated with radon implants, and they continued to be widely used until the early 1950s. Then, for reasons of dosage and cost, radon was super-seded by such artificially produced radioactive materials as cesium-137.

To summarize

Radon implants have been used widely and successfully for tumor-implant radiotherapy.

The advantage of this therapy is that the radiation that is designed to kill the tumor is deposited primarily in the tumor and not in healthy tissue.

In this chapter, we have touched only briefly on the effects of radon on uranium miners, who, until recently, breathed the gas in high concentrations. In fact, in recent years, these miners have been extensively studied in efforts to obtain some hard facts concerning just how much of a hazard radon presents. These studies have been fraught with difficulties, as we shall see in the next chapter.

Radon and Cancer

IN Chapter 3 we looked at how radiation damages human cells. All the information there essentially came from laboratory experiments with cells in test tubes, but what we want to know is whether radiation causes cancer in people. Assuming that it does, we would want to know what the chances are that a person exposed to a given amount of radiation will get cancer.

In fact, a great deal of our direct evidence on the relation between radiation and human cancer comes not from radon exposures, but from exposures of groups of people to different types of radiation. In the first part of this chapter, we will look briefly at the different groups of people who have been exposed to radiation and how many cancers these exposures actually caused. Then we will turn to the cancer effects that have actually been observed from groups of people exposed to radon—both miners and people living in high-radon areas.

Unfortunately, there have been several groups of people who have been exposed to radiations other than radon. The largest groups by far are the atomic-bomb survivors of Hiroshima and Nagasaki.

What problems do we face in studying these groups of people? In surveying a group of exposed individuals, we try to find two things: the amount of radiation absorbed by individuals and the increase in their cancer rate relative to a corresponding "control" group who received no extra radiation. With the information obtained, it should be possible to relate the amount of radiation received to the chances of getting cancer. Simple though it sounds, there are many problems involved.

First is the so-called latency period. Between the time of exposure and the time the cancer shows itself, there is a "latent" period during which the exposed person appears perfectly healthy. This span can be anywhere from one to 50 years (for radon-induced lung cancer it seems to range from about 10 to 40 years, with an average of about 20 years). Thus, exposed individuals must be monitored for their whole lives to see if a cancer appears. This makes for enormously long studies before definitive conclusions can be reached. An example of how shorter studies can mislead comes from the survivors of the atomic bombs dropped on the Japanese cities of Hiroshima and Nagasaki in 1945. The first reliable reports of cancer in survivors in these cities came about 10 years after the bombing. At this point, there were many more deaths from leukemia than from all other cancers combined. So it was thought at the time that leukemia was the prime hazard from radiation. As time has gone on, however, fewer and fewer new cases of leukemia have been reported in the survivors, whereas the rate for other cancers has increased (see Figure 6-1). What this means is that since leukemia has a shorter "latent" period, most of these cancers showed up early, whereas the other cancers have taken longer to appear. If, for example, this study had been stopped in 1955, 10 years after the bombing, completely misleading conclusions would have resulted.

The second difficulty in these human-cancer studies is the great uncertainties in estimating the actual radiation doses received. The people in Hiroshima and Nagasaki were certainly

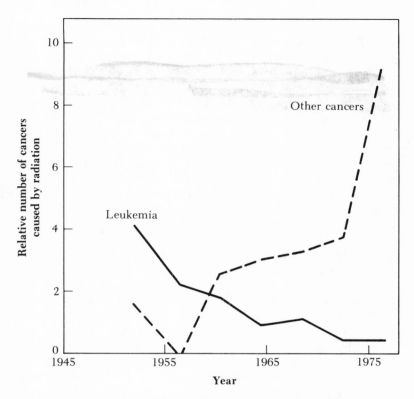

FIGURE 6-1 Relative numbers of excess cancers detected in survivors of the A-bomb at Hiroshima and Nagasaki, Japan. If the follow-up studies had been stopped in, say, 1954, the erroneous conclusion would have been reached that leukemia dominates over all other cancers put together. After a follow-up of over 30 years, however, it has become clear that this is not the case. (Adapted, with permission, from H. Kato and W. J. Schull, *Radiation Research* 90. Copyright 1982, Academic Press.)

not carrying dosemeters at the time of the explosions, so the best that can be done is to make rough estimates of their radiation doses. For the A-bomb survivors at Hiroshima and Nagasaki, scientists are still working on these estimates, over 40 years after the explosions. As we shall see, this problem is also present in studies of uranium miners: again, it is tremendously hard to reconstruct exactly the conditions at the bottom of a mine two or three decades ago.

The next problem relates to a "control" population. Cancer is a natural phenomenon. Even if there were no radiation, people would still get the disease. The way to look at radiation-induced cancer is to look at two similar groups, one who received radiation and one, the control group, who did not. Unfortunately, this is often very difficult. Consider again the survivors at Hiroshima and Nagasaki. The question is how to find a group of people with whom to compare cancer rates. One method might be to look at a group of people who lived outside these cities and therefore received no radiation. But the survivors were city dwellers and this other group would be country dwellers, and it is known that cancer rates are different in cities and in rural communities. The comparison would not be fair. What is actually done is to use a group of people from Hiroshima or Nagasaki who either lived on the very edge of the cities (and so received little radiation) or were outside the city on the day of the explosion. But this procedure may not be fair, either. After all, the irradiated group were survivors. They proved by surviving the atomic blast that they were physically tough. On the other hand, the group on the edge or outside the city were not selected for "toughness" in this way. Are these comparable groups?

The final problem has to do with statistics. Again, let us take Hiroshima and Nagasaki as our example. Out of a group of about 70,000 survivors studied, by 1985 about 5,900 of them had died of cancer. If they had not received radiation, about 5,600 might have been expected to die of cancer. The difference of 300 is clearly a significant one. But in any of the surveys of other irradiated groups (other than the A-bomb survivors), the actual number of people exposed is far smaller. The number of exposed people in any of the other studies might be a hundred times fewer. If all other conditions were the same as in Hiroshima and Nagasaki, let us see how the numbers would look if there were 100 times fewer people exposed. There would have been 700 (rather than 70,000) people in the exposed group, with only 59 getting cancer, as opposed to 56 on the basis of the control group. It would then be hard, on the basis of a difference of only 3, to say that people in the irradiated group were getting significantly more cancers than the controls. The more people in the exposed group, the more statistically reliable are the results. Of course, from a human point of view, the fewer people exposed, the better.

To summarize

Most information on the effects of radiation on humans comes from studies of groups of people who have been exposed to radiation.

Because of the latency period, individuals must be followed up throughout their lives to see whether they get cancer and at what age.

There are uncertainties as to the amount of radiation that individuals actually received.

There are difficulties in finding a "control" group, a group of people that is the same in all respects except exposure to radiation.

There are difficulties in drawing conclusions when the number of exposed individuals is small.

Having digressed on the difficulties of human studies of radiation-induced cancer, let us now look at a few of the main groups that have been exposed to radiation and what has been learned. As we have seen, by far the largest group of exposed individuals that has been studied are the survivors of the atomic bombings in Hiroshima and Nagasaki. For over 40 years, a team of Japanese and American scientists has been monitoring just under 100,000 survivors in the two cities. Every individual in the study has been individually interviewed and assessed concerning where they were at the moment of the explosion, whether they were indoors or outdoors, which way they were facing, whether they were standing, sitting, or lying down, and so on. With this information, an assessment can be made of the radiation dose to each individual. Over one third of survivors have died in the intervening years. For each person who dies, the cause of death is monitored and related to the radiation dose that the individual received. In this way, graphs like Figure 6-2 can be produced. Here the risk of getting cancer is related to the dose of radiation received. There are several things to be noted from this figure. First, at the lowest dose, there is still an increased risk because it is almost certain that any dose of radiation, no matter how small, can cause cancer. Second, as the dose increases, the risk of cancer increases.

Although the Japanese study provides the most reliable information that we have on the effects of radiation on humans, it still suffers from all the drawbacks that we have discussed. By far the most serious defect is the uncertainty in knowing the radiation doses to individuals. These doses depend on a variety of factors. First, we need to know the amount of radiation coming out of the bomb. In fact, determining this amount is difficult because the Hiroshima bomb was one of a kind—

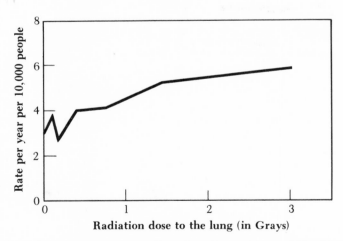

FIGURE 6-2   **Lung-cancer rate for A-bomb survivors at Hiroshima exposed to different doses of radiation to the lung. The cancer rate basically goes up as the radiation dose to the lung goes up (the "wobble" at very low doses is statistically insignificant). As a rough guide, living a lifetime in a house with 10 pCi/1 of radon would produce a radon dose to the lung of a little under 2 Gray. (Adapted, with permission, from Y. Shimuzu, et al., Radiation Effects Research Foundation Report TR12-87, 1987.)**

another one like it was never exploded in any of the subsequent A-bomb tests. Second, it is necessary to know the weather conditions, such as humidity and atmospheric pressure, because they would have determined how much radiation actually got through the air from the bomb to any given person. Third, as we saw, it is necessary to know where the person being studied was at the moment of the explosion. This problem is very significant because there are naturally many social and economic pressures preying on the memories of the survivors, such as avoiding the "stigma" of a high dose of radiation with its possible genetic consequences. Another factor might be that subsequent financial aid that each person received was related to the amount of radiation that person actually received.

With all these uncertainties, the Japanese data are still the best available. In the context of radon, we are particularly interested in the risk of lung cancer resulting from radiation. Based on the comparison with the control group, there were about 90 deaths from lung cancer that can be attributed to the radiation from the bombs.

The atomic-bomb survivors are by no means the only group of individuals exposed to excessive radiation. One other large category is people who received medical radiation therapy. The most important group in this regard are people who were suffering from a disease called ankylosing spondylitis. This disabling type of arthritis of the spine leads to almost complete stiffness of the back. In Britain, from 1935 to 1954, one successful type of treatment for this painful disease was irradiation of the spine with x rays. In all, over 14,000 people were given this treatment. By 1955, when it was realized that leukemia rates were rising in these patients, radiation therapy was discontinued. Since that time, major efforts have been made to follow up on as many of these patients as possible. In this case, an ideal control group is available, namely, people suffering from ankylosing spondylitis who did not receive radiation treatment. In fact, not only leukemia, but many other cancers, such as lung cancer, were found to be more frequent than the controls. On the basis of a comparison with the control group who had ankylosing spondylitis but did not receive radiation, of the 14,000 people who did receive radiation, about 400 cancer deaths can be attributed to the radiation. Of course, it is an arguable point whether, even if this had been known during the time the treatment was in fashion, it would not have been an acceptable risk anyway. How does a 3 percent risk of dying of cancer rate against lasting relief from almost continuous extreme pain?

Of the 400 excess cancer deaths resulting from the radiation, about 40, or one tenth, were from lung cancer, our particular concern here. It is interesting to compare the lung-cancer hazard, as estimated from the ankylosing spondylitis studies, with the estimate, described earlier, from the atomic-bomb survivors. In fact, in terms of excess lung-cancer deaths from radiation, the risk estimates from Japan are about 50 percent lower than those derived from the ankylosing spondylitis studies. (On the other hand, the corresponding leukemia risk estimates from Japan are two to three times higher — a good illustration of the uncertainties in these studies.)

To summarize

Other than miners, two main groups of people have been exposed to radiation that cause excess lung cancers: Japanese A-bomb survivors and British patients suffering from ankylosing spondylitis, a form of arthritis.

Estimates from these two groups of the risk of cancer from radiation differ significantly from each other.

The main sources of information on human lung cancer come from three studies of miners, conducted in North America and Europe over the past half-century. As we saw in an earlier chapter, it had been known since the sixteenth century that miners in the Erz mountains on the German-Czech border suffered from lung disease. In 1879 this lung disease was recognized as lung cancer — in miners on the German side of the border. It was not until 1926, however, that Czech miners were diagnosed as having lung cancer. This lack of communication was extraordinary, considering that the two mines are only about 20 miles apart. The lung-cancer rate in these miners was incredibly high. Anywhere from 30 to 75 percent of all miners died from this disease. Their death rate was estimated to be about 50 times higher than that of a non-mining population.

The cause of this huge rate of lung cancer was debated for many years. For a long time, the most popular theory was that the cancer came from breathing in dust or perhaps toxic metals such as arsenic. Another popular idea was that because inbreeding was common in the mining communities in Germany and Czechoslovakia, the miners were showing a genetic susceptibility to lung cancer. On the other hand, workers in similar mines with all these same characteristics (dust, toxic metals, an inbred population) *except* radioactivity were not showing excessive lung-cancer rates. Thus, radon became a prime candidate as the cause of the excess lung cancers.

Finally, by the early 1950s, it became clear that inhaled radon daughters were a cause of the lung cancers in miners, and soon after, studies of these miners began. The idea was to try to relate lung-cancer rates to the various factors that might be important, such as the amount of radon daughters inhaled and the miners' smoking habits.

In fact, there are about 20 different mining areas around the world where miners have been exposed to large amounts of radon daughters; however, only three have enough miners and have been studied in enough detail to be useful for our purposes. These three regions are in Colorado, Czechoslovakia, and the Canadian province of Ontario. We will take a brief look at each of these, bearing in mind the questions that we discussed earlier in the chapter: has the study been going long enough for all the cancers to show up? How well can the amount of radon exposure be estimated? Is there a suitable "control" group? Are there enough people in the study?

THE COLORADO STUDY

At the end of World War II, demand for uranium dramatically increased, first for nuclear weapons and then for atomic power stations. The main source of uranium in the United States was the Colorado plateau, a mountain area of southwestern Colorado and southern Utah. In 1949, because of the increasing concern of the lung-cancer hazards that faced the uranium miners, the U.S. Public Health Survey started a study to determine the number of miners who were contracting lung cancer and also the amount of radon-daughter exposure that the miners were receiving. This study, in one form or another, has continued to the present day. Over the years, as the awareness of the lung-cancer hazard has increased, various regulations have been put into place to limit the radon concentration in the air in mines. Figure 6-3 shows the dramatic fall-off in average radon concentration in U.S. uranium mines in the post-war period.

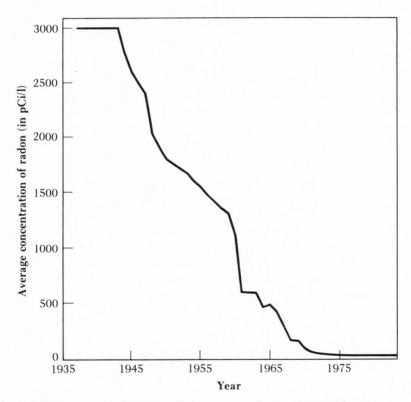

FIGURE 6-3   Average radon concentration in U.S. underground uranium mines from 1937 to 1980. As awareness of the hazards of radon increased and steps were taken to improve ventilation, the levels fell dramatically. (Adapted with permission from L.W. Swent, "Radiation Hazards in Mining: Control, Measurements, and Medical Aspects," M. Gomez, Editor. Copyright © 1982, Society of Mining Engineers, New York.)

The numbers in Figure 6-3 are estimates based on measurements actually taken inside mines. But in the 1950s, these measurements were few and far between. In addition, two groups of people were making measurements, each likely to have their own biases. The first group were the mining companies themselves, whose natural tendency would be to underestimate radon levels. Many of the later measurements, however, were taken by government agencies which, if they were to err at all, would be more likely to overestimate the radon levels.

Because so few measurements of radon levels were actually taken, educated guesses have been made to fill in the gaps. In fact, only about one-tenth of the miners in the Colorado study had exposure levels estimated directly from measurements. For the rest, exposure levels were based on some "guestimation" procedure. Some figures from this and other studies are shown in Table 6-1. Although the number of miners in the Colorado study is quite large (about 3,000), recall that only around one in ten miners have their radon exposures known with any de-

TABLE 6-1 *Some details of the three main studies of uranium miners exposed to high levels of radon*

	Colorado, U.S.	Czechoslovakia	Ontario, Canada
Number of miners	2975	4043	10661
Average time in study (years)	25	21	19
Percentage dead at end of study	33	35	9
Average radon exposure in WLM	509	227	65
Number of lung cancers found	157	484	82
Number of lung cancers expected, based on controls	49	98	57
Excess number of lung cancers	108	386	25

gree of certainty. We should also note that only about one-third of the miners in the study have so far died, so it is not known what fraction of the rest will die of lung cancer.

Over 70 percent of the Colorado miners are smokers. This is about twice the average in the United States. So, if the results from the miners are to be useful for the general population, the effects of smoking must be taken into account. We will return to this problem later in the chapter.

Another problem in relating the information from mines to the general environmental situation involves the amount of radon daughters to which the miners were exposed: the average radon-daughter exposure of the miners is 509 WLM (Working Level Months, see Chapter 4). The average lifetime exposure for people living in houses is probably less than 30 WLM. Before the information from mines is useful for risk estimates in environmental situations, the risk at these high levels needs to be related to the risk at much lower levels. But it should be kept in mind that people living their lives in houses with very high radon levels may well get exposures comparable to those which the miners received.

Finally, let us look at the actual lung-cancer deaths in the study, shown in Table 6-1. In the period 1951 to 1982, there were 157 deaths from lung cancer, whereas from the "control" group, who were not exposed to radiation but had comparable smoking habits, only 49 deaths would have been expected. In other words, there was a significant and dramatic increase in the lung-cancer rate due to radon daughters. To illustrate the effects of lower levels of radon, Figure 6-4 breaks down the lung-cancer statistics according to the actual estimated radon-daughter exposures that the miners received. In the group of miners who received the lowest exposures (less than 120 WLM), there were actually the same number of cancers in the exposed group as in the control group who received no radiation. At first sight, this comparison seems to tell us that a small amount of radiation is not harmful at all, but in fact it is more likely a case of insufficient follow-up. As we saw in Figure 6-3, the exposure levels in miners went down dramatically between about 1950 and 1970. This decrease means that most of the miners who did receive the lower radon exposures worked in the mines at a later time, compared with the rest of the miners in the study. So the time that has elapsed since their going

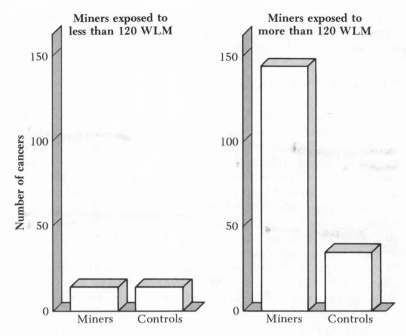

FIGURE 6-4   Number of lung-cancer cases observed in Colorado uranium miners between 1950 and 1983. The miners are compared here with a control group not exposed to excess levels of radon. For miners who received high radon exposures (more than 120 WLM), a larger increase in the number of lung cancers observed is visible. For miners who received lower exposures, no significant increase has yet been observed, partly because this group has not been observed for a long enough period of time. As a rough guide, 120 WLM is the exposure that a person would receive living a lifetime in a house with about 8 pCi/l of radon. Note that miners getting extremely high exposures (more than 2,000 WLM) have been excluded from this figure because this situation would be very rare in a house. (Data adapted with permission from "The Health Risks of Radon and other Internally Deposited Alpha Emitters." Copyright © 1988, National Academy Press, Washington, D.C.)

down the mine is less than average. As we saw, lung cancer takes, on average, 20 years to show up, so many more "potential" lung cancers have probably not yet shown themselves in the miners with lower exposures. The conclusion is simply this:

based on the results, it is hard to estimate lung-cancer rates for people exposed to radon levels in houses, who typically receive less than 30 WLM over a lifetime.

THE CZECH MINERS

Mining in the mountains near the Czech-German border has been going on for many hundreds of years. Over 4,000 miners who have worked since 1948 in the uranium mines have been studied by Czech scientists. Apparently, many more measurements of radon levels in these mines were taken than in the U.S. study, which should lead to more reliable estimates of radon exposure. However, it has been pointed out that for a given mine in a given year, the measurements are all remarkably similar to each other. Given all the factors that affect radon levels in mines, such as ventilation and moisture levels, it would be very surprising if daily measurements over a period of a year all gave very similar results. More likely would be a considerable variation among these measurements; the fact that they are all so similar casts some doubt on their credibility.

Looking at Table 6-1, where all the facts and figures for the Czech study are listed, you can see that perhaps the biggest difference between this and the U.S. study is the average radon-daughter exposure that the miners received: 227 WLM, compared with 509 for the U.S. study. While this is still higher than exposures that most people in houses might receive (typically less than 30 WLM), it is certainly much closer. As in the U.S. study, we find a large excess of lung cancers: 484, compared with a number (based on the controls) of 98.

ONTARIO URANIUM MINERS

Uranium mining in Ontario, Canada, started in 1955. Because this study began a decade later than in the other two countries, much more attention was paid to measurements of radon levels in mines. Up to 1981, 131,000 measurements of radon-daughter levels were taken in the Ontario mines, far more than in the U.S. study. This figure corresponds to about one measurement in every mine every day. As might be expected, these measurements vary considerably from day to day and from month to month, because of the ventilation and moisture conditions. These variations make it hard to estimate exposures for individual miners. Most of the exposure estimates probably have enough uncertainty associated with them that they could easily be twice as low or twice as high as actually estimated.

One complicating factor is that many of the uranium miners were previously employed as gold miners — another occupation with a high risk of lung cancer. This group has been excluded from the figures in Table 6-1. Another difficulty with this study is that because the mining started later than for the other studies, the follow-up period is shorter. Up to 1981, less than one in ten of the Ontario miners had actually died, so there are certainly going to be many more lung cancers that have not yet appeared.

The most interesting figure in Table 6-1 for the Ontario miners is the average exposure of 65 WLM, which is far lower than either the American or the Czech study. This level really does approach exposures that people could receive from living in houses. From Table 6-1, up to 1981, 82 lung-cancer cases had been observed among the Ontario miners, compared with a number of 57 based on corresponding controls. But, as we saw, not all of the lung cancers have as yet shown up, because of the short follow-up time.

Having looked at the three major studies, the obvious question is whether the observed lung-cancer hazards are similar from study to study. We cannot directly compare the excess cancers observed in the different studies, for several reasons. First, the average radon exposures are different (see Table 6-1) in these different studies, and we have seen that as the radon exposure goes up, the lung-cancer risk goes up as well. We can get around this by comparing the excess lung cancers for the same exposure, say, 1 WLM. For example, the excess number of lung cancers for the Colorado miners is 108 (that is, 157, the observed number, minus 49, based on the controls). This gives us 108 lung-cancer cases for an average exposure of 509 WLM, which is about 0.2 cases for 1 WLM. Next, we have to standardize the number of people exposed. The Colorado study was for 2,975 exposed people. To compare this figure with numbers in the other studies, we will standardize each to, say, one million persons. For the Colorado study, then, we get about 70 lung-cancer deaths per WLM per million people. Finally, we have to standardize the number of years the miners are in the study because the more years the miners are studied, the more lung cancers we will see. From Table 6-1, the average number of years that a miner has been in the Colorado study is 22, so the final risk is 70 divided by 22, or about three lung-cancer deaths per year per million people exposed to 1 WLM. Doing the same arithmetic for the Ontario miner results, we get about 2.5 lung-cancer deaths per year per million people exposed to 1 WLM, but for the Czech miners, about 20.

Our results, then, are by no means consistent. Yet these are the main results on which all our risk estimates are based. Which are we to believe? As we have seen, there are problems with estimating the exposures in all the studies, so none can really be ruled out on this basis. Smoking habits are quite similar among miners in the different studies, so this factor cannot account for the difference. One difference between the studies is the length of follow-up. As we saw, the latent period —the time between radiation exposure and onset of cancer— is typically about 20 years for lung cancer. Because the follow-up times are a little different from each other, a reasonable expectation would be that the Colorado and Czech studies would show the largest lung-cancer rates (as they have the longest follow-up times for cancers to show themselves), with the Ontario rates somewhat lower. However, the Czech study seems way out of line with the other two and must be considered the "odd-man out."

To summarize

Three main groups of uranium miners—from Colorado, Czechoslovakia, and Ontario—have been studied to provide risk estimates for people exposed to radon daughters.

The three groups differ significantly in terms of the average radon exposure that they received, as well as in the numbers of years that the miners have been studied.

Risk estimates for the three groups are not consistent with each other: the risk estimates from the Czech miners are much higher than either of the others.

SURVEYS AMONG THE GENERAL PUBLIC

The studies of uranium miners are not providing exactly the information that we want: the exposures are too high, smoking is an interfering problem, and the results are inconsistent. It might seem logical, then, to survey groups of the general public who are exposed to high natural levels of radon in their homes, and see if an excess can be detected there. In fact, such surveys are extremely hard to do, largely because of a problem we discussed earlier: statistics. The excess number of lung cancers is related to exposure, which is much smaller for people living in houses, relative to people working in mines. So there will be a much smaller rate of excess cancers than we saw with the miners, who received the larger exposures. If the average radon exposure in houses were, for example, ten times lower than the average miner exposure, we would need ten times more people in a survey of houses than in the miner studies, say 50,000 people. Ideally, we would have to interview each of these 50,000 people and take measurements in their houses, to estimate their radon-daughter exposure, as well as their smoking habits. Then this group would need to be followed up for the rest of their lives to relate the number of lung cancers showing up to radon exposures.

No full-scale study of this kind has yet been undertaken, but several Scandinavian countries have tried scaled-down versions. In Finland, a survey was done of 60,000 residents of rural areas where radon levels averaged about four times the national average. No difference in lung-cancer rates could be detected between this group and the national average. On the other hand, however, only a few of the houses of the 60,000 subjects were surveyed for radon levels, so it is not certain whether this group really did have higher radon exposures. A similar study of 10,000 homes in Canada was equally inconclusive. Both surveys did not take into account individuals' smoking habits, which makes it difficult to draw any conclusions. One study in Norway has tried to take into account the effects of smoking so that the effects of radon alone can be studied. This study did find that the risk of lung cancer from radon increased for people living in houses with high radon levels.

In the United States, several studies have examined lung-cancer rates of people living in the Reading Prong, a geological area (see Figure 4-8) known to have high levels of radon in the quarter of a million homes built on it. Probably over half these homes have radon concentrations over 4 pCi/l (compared with one in 12 homes for the United States as a whole). Lung-cancer rates for people living in this area from 1950 to 1979 are illustrated in Figure 6-5, for counties actually in the Reading Prong, for counties just on the edge, and for counties not on the Reading Prong. There is a just-visible increase for people actually living on the Reading Prong. This study did not examine radon exposures to individuals; it just compared lung-cancer rates county by county. A much more detailed study has been started of lung cancers in women in central and eastern Pennsylvania; however, it will be many years before any definitive results come out of these studies.

Overall, it simply has not been possible to get actual estimates of risk from environmental studies of the general public —the uncertainties are just too large. This is not to say that it is impossible to get risk estimates out of such studies, but the surveys would need to be much larger and more detailed than the current ones.

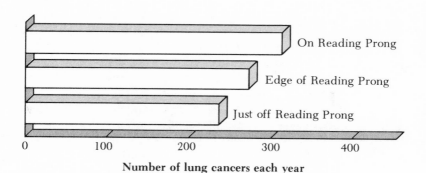

Number of lung cancers each year
(per million people)

FIGURE 6-5   Results of a survey of lung-cancer rates for people living either in, or on the edge of, or just off the "Reading Prong" geologic region illustrated in Figure 4-8. There is a significantly larger number of lung cancers for people living on the Reading Prong, where average indoor radon levels are considerably elevated. (Adapted with permission fron V.E. Archer, *Archives of Environmental Health* 42. Copyright 1987, Helen Dwight Reid Educational Foundation.)

To summarize

Ideally, risk estimates for people exposed to environmental levels of radon would come from lung-cancer surveys of people living in high-radon areas.

As of 1988, there is only marginal direct evidence suggesting that people living in high-radon areas have higher lung-cancer levels. This lack of evidence is probably not because the danger is not present, but because appropriate large-scale surveys have not yet been undertaken.

We are forced to rely on the results from the surveys of miners because we simply have nothing else. But these risk estimates from miners cannot be used directly to give risk estimates for the general public, for reasons we have discussed. First, the miner populations have not been studied until they are all dead, so we do not know how many more lung cancers will show up. Second, all the miners are adult males, and we clearly also need information about women and children. Third, the problem of smoking confuses the results. Seventy percent of miners smoke, but, in the United States, less than half of the general public smoke. Smoking certainly causes lung cancer, so how can we subtract out the effects of smoking to focus on the effects of the radon?

Currently, there is one method that is most often used to estimate risks from exposure to radon in homes, based on the results from miners. This is the so-called "relative-risk" method, where the risk from radon is related to the risk of lung cancer that a person would have due to all causes (including smoking) other than radon. These rates from nonradon-related causes are shown as the lower lines in Figure 6-6, and are much larger for men than for women, primarily because men smoke more than women. Let us now assume that exposure to some level of radon increases the lung-cancer rate by 20 percent. Then the new, increased rate is shown by the upper dashed curves in Figure 6-6. The difference between the two curves (the shaded region in the figure) would then be due to radon alone. Note that this shaded region is much larger for men than for women because, in absolute terms, a 20 percent increase on

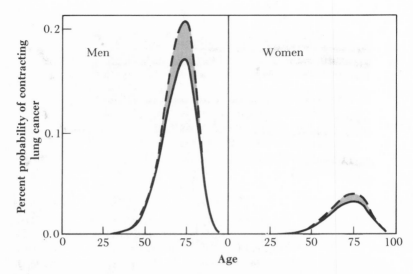

FIGURE 6-6 The "relative" approach to risk estimation. Lung-cancer rates due to causes other than radon are shown by the lower full lines; the rates, if they are increased (because of radon) by, say, 20 percent, are shown by the upper dashed lines. The difference (shaded) will then be due to radon alone. Because the nonradon-related rate is larger for men than for women, the shaded area (due to radon) will also be larger.

a large rate (i.e., for men) is bigger than a 20 percent increase on a small rate (i.e., for women).

This same idea of "relative risks" would also apply to smokers who, of course, have a much higher lung-cancer rate than nonsmokers. A 20 percent increase in the lung-cancer rate for smokers would, in absolute terms, be a far larger increase than a 20 percent increase for smokers. If this is true, then many lung cancers are caused not by cigarettes or radon by themselves, but by the combination of both. This conclusion is very much in accord with the ideas of the causes of cancer described in Chapter 3 (see Figure 3-4). In general, then, the relative-risk approach predicts that radon will be more dangerous for smokers and for men (because they have high lung-cancer rates anyway) than for nonsmokers and for women.

The useful feature about this "relative risk" approach is that it enables us to extend our risk estimates beyond the limits of the miner data. For example, knowing the percentage increase in lung cancer from radon and the lung-cancer rate for women, we can make radon risk estimates for women. Also, because we have information about lung-cancer incidence at all ages, we can extend the information from the miners to all ages, and thus get lifetime risk estimates.

A problem of using this model — or any model based on the miner data — is that we do not know the correct relative risks to use. In terms of these relative risks — the percentage increase in lung-cancer risk over the natural rate without radon — the three miner studies give different answers. The Colorado study shows about a 0.5 percent increase in risk for each WLM of exposure to radon daughters; the Ontario study shows about 1 percent. But the Czech study gives almost 2 percent, much higher than the other results. A reasonable compromise might be 1 percent, but the real answer could easily be twice as high or twice as low.

A more basic problem with this relative-risk approach is that we really do not know if it is correct. There exist other, quite plausible approaches in which the excess cancer incidence is not related to the lung-cancer rate, but is simply a constant amount. The miner information (and also the information from the Japanese A-bomb survivors) is generally consistent with the "relative risk" approach that we discussed above, but it is very hard to rule out other possibilities.

To summarize

It is important to be able to extend the risk estimates from miners to other situations (for example, lower exposures, women, non-smokers, lifetime risks).

A reasonable technique is to assume that exposure to radon increases the lung-cancer risk by some percentage, depending on the exposure. This method is called the "relative-risk" approach.

The relative-risk approach suggests that people who are already at a high risk of getting lung cancer (men and/or smokers) will, in absolute terms, be at a bigger risk from radon than people who have low lung-cancer rates (women and/or non smokers).

The effects of radon and smoking together are probably such that their combined effect is larger than the sum of their individual effects.

7

The Hazards of
Radon in Houses

FINALLY, in this chapter, we estimate the risks associated with living in houses containing radon. We will look first at the risks to the population as a whole to determine how many deaths per year can be blamed on radon. After that, we will look at the risk that an individual person faces by living in a house containing a particular concentration of radon.

From Chapter 6, it is clear that we do not know the risks very well. A reasonable approach therefore is first to estimate the risks while being as pessimistic as possible about the hazards of radon. Then, we make risk estimates at the other extreme, being as optimistic as possible. The true risk is going to be somewhere in between. We will have three sets of risk estimates: the pessimist's set, which will have the highest estimates of the risk; the optimist's set, which will have the lower risk estimate; and the "reasonable person's" set, which will be somewhere in the middle.

First, let us estimate how many people will die each year in the United States because of exposure to radon at home. We will need two important pieces of information: the average radon concentration in a house in the United States and the risk of lung cancer for a given radon exposure. The average indoor radon concentration, as measured in the two major U.S. studies (see Table 4-1), is in the region of 1.5 to 3 pCi/l. Because of biases in the surveys that we discussed in Chapter 4, a reasonable estimate is probably 1 pCi/l. The optimistic estimate is 0.5; the pessimistic estimate, 2.0. In terms of lifetime radon-daughter exposure, these numbers correspond to about 15 WLM, 7.5 WLM, and 30 WLM.

Next we need to know the increased risk of lung cancer for each WLM of radon-daughter exposure. Recall that the Czech study gave about 2 percent per WLM, whereas the Canadian and U.S. studies gave about 0.5 percent per WLM. A reasonable estimate, therefore, seems to be 1 percent, with optimistic and pessimistic estimates of 0.5 percent and 2 percent, respectively. (We might note here, incidentally, that the EPA has estimated the risk at somewhere between 1 and 4 percent per WLM, which is rather higher than most other estimates).

To produce actual risk estimates requires a small amount of math, which we will not go into here. The information needed for the calculation is similar to that shown in Figure 6-6. In particular, we need the chances of dying of lung cancer and the chances of dying for any other reason, for people who live to be any given age. This information also needs to be broken down into men, women, smokers, and nonsmokers.

The results are shown in Figure 7-1 for "reasonable people," for optimists, and for pessimists. The total number of people who die each year of lung cancer in the United States is about 139,000; of these, about 16,000 or about one in nine of all lung cancers (the "reasonable" estimate) can be related to radon, although the proportion could be as high as one in three (pessimistic estimate), or as low as one in thirty (optimistic estimate).

This predicted annual death rate is for everyone in the United States, men and women. If we break the number down

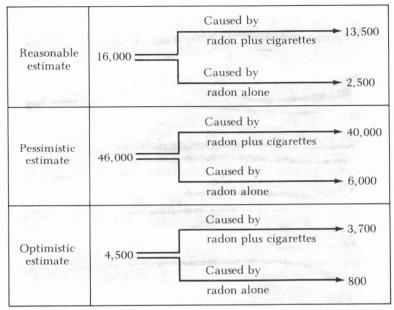

FIGURE 7-1 Estimates of the number of lung-cancer deaths caused each year by radon in the United States. The reasonable estimates are our best estimates, but the true numbers could be as high as the pessimistic estimates, or as low as the optimistic estimates. Note that most radon deaths, according to the "relative-risk" model, are due to the effects of radon *and* smoking.

by sex, the prediction is that about 12,000 men and 4000 women will die each year because of radon. This big difference between men and women is not because men inhale any more radon daughters than women, but is because the "natural" (i.e. without radon) lung cancer rate for men is much higher than for women. Now, our relative-risk approach (see Chapter 6) says that an exposure of radon will increase the risk by some proportion (for example an extra 10 percent) over the lung-cancer risk that is already present. An extra 10 percent of a big risk (lung cancer in men) is, in absolute terms, bigger than an extra 10 percent of a smaller risk (lung cancer in women). Thus radon will produce far more deaths in men than in women.

To summarize

The average radon concentration in houses in the United States is probably in the range of 0.5 to 2 pCi/1. Our best estimate is 1 pCi/1.

The increased risk per exposure to 1 WLM of radon daughters is probably in the range of 0.5 to 2 percent. Our best estimate is 1 percent.

Based on these figures, the number of deaths per year that can be attributed to radon in homes is between 4,500 and 46,000. Our best estimate is 16,000, or one in nine of all lung-cancer deaths.

Of the predicted 16,000 radon-related deaths each year in the United States, about 12,000 will be men and 4,000 will be women.

Of the predicted 16,000 radon-related deaths each year in the United States, about 13,500 will be smokers and 2,500 nonsmokers.

So far, we have looked at the effects of radon on the whole U.S. population. Of more immediate concern is the effect of radon on individual people: this effect will depend, of course, on the amount of radon in the particular individual's house. Let us consider four families, spending their lives in four houses with low, average, high, and very high concentrations of radon. To cover all the possibilities, let us suppose that in each house there is a man and a woman who smoke an "average" number of cigarettes, and a man and a woman who do not smoke. What are the chances any of these people will die of lung cancer caused by radon? The answers are shown in Figures 7-2 through 7-5, in which the "reasonable" relative risk of 1 percent per WLM has been assumed. To take, for example, the most dramatic case (Figure 7-5), a male smoker living in a house that has 20 pCi/1 of radon has a one in four chance of dying of radon-produced lung cancer. As we saw in Chapter 4, there are probably about 150,000 houses in the United States with 20 pCi/1 or more of radon.

What is very striking about Figures 7-2, 7-3, 7-4, and 7-5 is the dramatic effects of sex and of smoking habits. A man is about twice as likely as a woman to contract lung cancer from

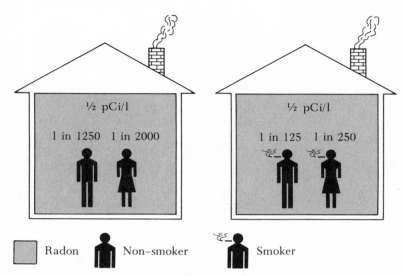

FIGURE 7-2 Estimate of the chances of dying of radon-produced lung cancer for people spending their lives in houses containing ½ pCi/1 of radon. Estimates are shown (left to right) for male and female nonsmokers, and for male and female smokers. The estimate assumes that each person spends three quarters of his or her life in the house.

radon, and an average smoker is ten times more likely to get lung cancer from radon as a nonsmoker. The effect of smoking is illustrated in Figure 7-6 for a smoking and a nonsmoking couple in identical houses, each containing 4 pCi/1 of radon. The chances of either of the nonsmokers dying of radon-produced lung cancer are only about 1 in 200, whereas the chances of one of the smokers dying of lung cancer caused by the same concentration of radon is estimated to be 1 in 20. Of course the risks would be correspondingly lower for a light smoker and greater for a very heavy smoker. The risks quoted here are for an "average" smoker, one who smokes ten to twenty cigarettes each day. Someone who smokes twice as much would almost double the risk of getting lung cancer from radon.

The results quoted in this chapter depend a lot on the method of risk estimation that we use. As we saw in the last

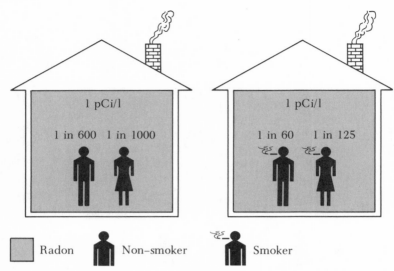

FIGURE 7-3 Chances of dying of radon-produced lung cancer, as in Figure 7-2, for occupants of houses containing 1 pCi/1 of radon.

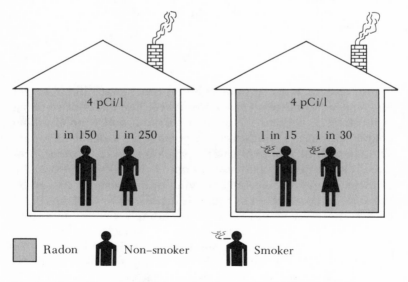

FIGURE 7-4 Chances of dying of radon-produced lung cancer, as in Figure 7-2, for occupants of houses containing 4 pCi/1 of radon.

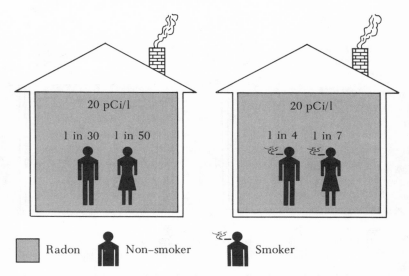

FIGURE 7-5 Chances of dying of radon-produced lung cancer, as in Figure 7-2, for occupants of houses containing 20 pCi/l of radon.

chapter, the problem is to generalize from information about miners to the rather different situation of people in houses. All the results quoted here were made with the "relative-risk" approach, described in the last chapter: it is the opinion of most scientists that this is the most appropriate method, but there are other approaches that we cannot rule out. To illustrate that the use of different methods of risk estimation really does make a difference, consider the lifetime risk from being in a house having a radon concentration of 4 pCi/l. If we average the lung-cancer risks in Figure 7-4 over both sexes and also over smokers and nonsmokers, we get an overall risk of lung cancer of about 1 in 50 (see Figure 7-7). In estimating this same risk, the National Council on Radiation Protection and Measurements analyzed essentially the same miner information but used a different type of approach from the relative-risk method used in this book: they estimated an overall lung-cancer risk of about 1 in 100 — twice as low as that predicted by our relative-risk approach.

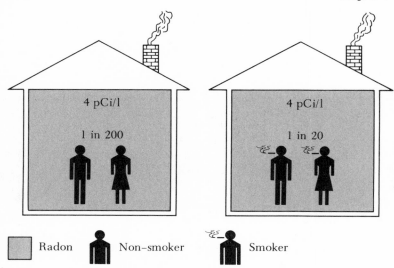

FIGURE 7-6   Estimate of the chances that an individual will die of radon-produced lung cancer for people spending their lives in houses containing 4 pCi/l of radon. The risks are given separately for a nonsmoking and for a smoking couple. The estimate assumes that each couple spends about three quarters of its time in the house.

Finally, then, we have our answer: people living their whole lives in a house containing, say, 4 pCi/l of radon will on average have a 1 in 50 chance of dying of radon-induced lung cancer. If the occupants are nonsmokers, the risk is down to about 1 in 200; if the occupants are smokers, the risk is up to 1 in 20.

To summarize

For people living their whole lives in an "average" house containing 1 pCi/l of radon in the air, estimates of the lifetime risk of dying from lung cancer are

   1 in 500, male, nonsmoker

   1 in 60, male, smoker

   1 in 1,000, female, nonsmoker

   1 in 125, female, smoker

   1 in 200, average for everyone

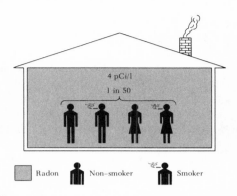

FIGURE 7-7 Estimate of the chances that an individual will die of radon-produced lung cancer for people spending their lives in houses containing 4 pCi/1 of radon. The risk shown here is averaged over men, women, smokers, and nonsmokers.

As the radon level in the house goes up, the risk goes up: the estimated risk of dying of lung cancer for a male smoker living in a house containing 20 pCi/1 of radon is as high as 1 in 4.

The risk estimates depend to some extent on the method used to generalize from miners to people in houses: estimated risks might be as much as twice as low if a different technique is used.

How can we use this information? In practice, we use it to decide on an *action level*: if a house has a radon concentration above this action level, some action should be considered to reduce the levels in that house.

This procedure is not really very logical, because *any* level of radon in a house will increase the occupant's chances of getting lung cancer. The ideal approach would be to use a principle called ALARA ("As Low As Reasonably Achievable"): quite simply, everyone should be exposed to the smallest radiation dose that is reasonably possible. No unnecessary exposure should be allowed. According to this principle, it is desirable to reduce the radon level in the U.S. housing stock as a whole, not just in houses with high levels. As a first step, however, towards controlling the problem, a practical procedure is to decide on some radon concentration — the action level — and try to bring all houses below this level.

What level should we choose? The choice is really somewhat arbitrary. The U.S. National Council on Radiation Protection has suggested that a risk of more than 1 in 50 of getting lung cancer from radon should be avoided. From Figure 7-7, 1 in 50 corresponds to a radon level of about 4 pCi/1, which is indeed the action level recommended by the EPA.

According to this standard, if a house has a radon concentration above 4 pCi/1, it is said to be in need of fixing or

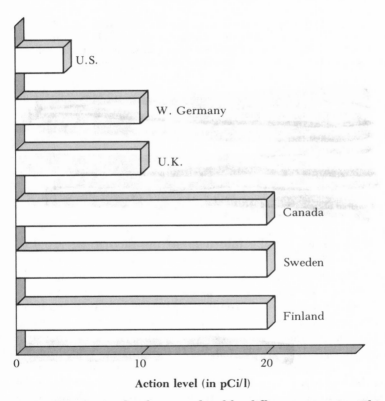

Action level (in pCi/l)

FIGURE 7-8 Action levels as regulated by different countries. This term refers to the radon concentration in a house above which it is suggested that the homeowner consider taking some remedial measures.

mitigating to reduce its radon level. Actually, different countries have suggested different action levels, as shown in Figure 7-8. The United States has the most stringent action level in the world, but probably the most realistic. According to Figure 4-13, one in twelve of all homes — about 6 million — are above this action level of 4 pCi/1.

To summarize

All levels of radon in houses are hazardous to some extent.

The action level specifies the radon concentration below which no action is recommended and above which some remedial measures are called for.

The U.S. Environmental Protection Agency has suggested an action level, 4 pCi/1, that seems reasonable.

From a practical point of view homeowners will decide on their course of action according to whether or not their home has radon concentrations above the action level. In the next chapter, we will look at the first step a householder needs to take: measuring the radon level in the house.

Measuring the Radon Concentration in Your Home

W E now turn to the problem of estimating the radon risk from living in your own home. As we have seen, the risk of lung cancer is directly related to the concentration of radon in the home: the larger the radon concentration, the larger is the risk. So it is essential to get an estimate of how much radon is in your house. Fortunately, there are simple, inexpensive ways of surveying a house to find its approximate radon concentration. In this chapter we will discuss exactly how to take a radon measurement in a home.

The first thing to be aware of is that radon levels in any given house vary considerably with time. Figure 8-1 shows how radon levels typically vary in a house over the seasons. The lowest levels are to be found in the middle of the summer, when most windows are open, and the highest levels are to be found in the middle of winter, when the house is tightly closed

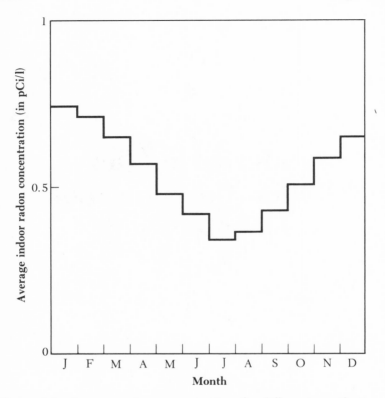

FIGURE 8-1   Average radon concentrations for different months in the year for ground-floor living rooms. These results were taken from a survey of 2,000 British houses. (Adapted with permission from J.C.H. Miles and R.A. Algar, *Journal of Radiological Protection* 8. Copyright © 1988, IOP Publishing Ltd.)

and also much warmer than outside. To be on the conservative side, we should take measurements either in winter or over a whole year.

Day to day, radon levels can fluctuate a great deal. Figure 8-2 shows the results of measurements taken every twelve hours in a house in England, over a period of ten days. Dramatic day-to-day variations can be caused by such things as opening and closing of windows and by changes in climatic conditions, such as wind, temperature, and pressure. The average in the particular house in Figure 8-2 was 56 pCi/l, but the

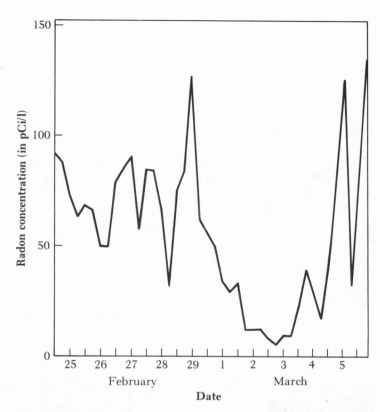

FIGURE 8-2 Actual variations in radon concentrations in the living room of a house. Measurements were taken over 6-hour periods of time for a total of 10 days. These variations are very hard to predict because they are caused by a variety of factors such as wind, temperature, and pressure. (Adapted with permission from K.D. Cliff, et al. in "Radon and Its Decay Products," Philip Hopke, Editor. Copyright © 1987, American Chemical Society.)

individual measurements ran from 6 to 140 pCi/1. What this means is that measurements taken over a few hours or a day are of little use. At the very least, a measurement needs to be taken over a span of several days.

Let us now look at the different techniques that can be used to make these measurements. In fact, there are two devices that are practical for household use. One is the charcoal canister and the other is the alpha-track detector. They are each useful in different circumstances; both involve exposing the device to air in the home and then sending it out to a laboratory to have the results processed. Let us look at them both in turn.

THE CHARCOAL CANISTER

The charcoal canister (see Figure 8-3) is an open tin can, a few inches across, containing about an ounce of a type of powdered charcoal called activated charcoal. The charcoal is kept in place by a fine wire mesh. On top of this mesh is a thin nylon screen. The can is kept sealed until the beginning of the measurement, and then opened. When the can is opened, radon in the air can diffuse through the nylon screen and will be adsorbed on the charcoal. After a few days, the can is sealed up again, which prevents any more radon from getting in. At this point, the charcoal has adsorbed an amount of radon that is a measure of the concentration of radon in the air while the can was open.

FIGURE 8-3 Typical charcoal-canister radon detector. The can, in this case about 3 inches across, has been opened up to show activated charcoal and the wire mesh to keep it in place.

The radon adsorbed in the charcoal is radioactive. If the charcoal is put on a specialized gamma-ray detector, measurement of the radiation coming out of the charcoal gives an estimate of how much radon is in the charcoal and, hence, how much radon was in the air while the can was open.

The specialized gamma-ray detector costs anywhere from six to ten thousand dollars and requires a trained operator. The usual procedure, therefore, is to purchase the canister through the mail, open it up in the house for a few (two to seven) days, and then mail it back to a laboratory for processing. It is important to send it back as soon as possible, because radon-222 has a half-life of only a few days, so the radon will be rapidly decaying away.

The charcoal canister is, by contrast, inexpensive — the component parts cost less than a dollar. The charcoal can even be reused, once it is heated up to get rid of the radon. The system works slightly less well when the air is humid, because the charcoal does not adsorb well when it is damp, but this is not a major problem. It is also extremely easy to use: the can is simply opened up for a few days, closed up again, and mailed out for processing.

The charcoal-canister system is not useful for taking measurements for longer than a week. The reason is that radon adsorbed on the charcoal at the beginning of the week probably will have decayed away by the end of the week, and so will not be measurable when the canister is sent to the laboratory for analysis. This is both an advantage and a disadvantage. The advantage is that the measurement is short, so that the homeowner gets the answer within, say, two weeks of deciding to take a measurement. The disadvantage of the short measurement period is that it means that the measurement is subject to the sort of fluctuations that are shown in Figure 8-2. For example, you can see from the example in Figure 8-2 that if a charcoal-canister measurement had been taken starting on February 25 and ending on February 29, the results would be far higher than from a charcoal-canister measurement starting on February 29 and ending on March 4.

To summarize

The charcoal-canister radon detector is inexpensive and easy to use.

Because the charcoal-canister detector measures radon levels only over a few days, it is affected by the large fluctuations in indoor radon levels from day to day and from season to season.

The charcoal canister is ideal for a first radon screening measurement.

THE ALPHA-TRACK DETECTOR

The second measuring system complements the first in that it is designed to sit in the home for much longer periods, anywhere from three months to a year. As with the charcoal canister, the procedure is to purchase the detector by mail, expose it to the air, and then send it back for processing. The detector (see Figure 8-4) consists of a thin piece of a special type of

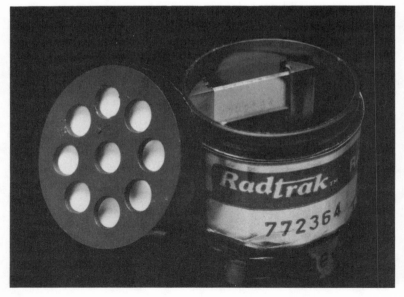

FIGURE 8-4 Typical alpha-track radon detector. The can, in this case about 1 inch across, has been opened up to show the piece of sensitive plastic inside. (Courtesy of Terradex Radon Detection Products, Illinois.)

plastic placed inside a small can, often the same size as the charcoal-canister can. Above the can is a thin filter, and when it is not actually measuring radon, the can is sealed with an air-tight top.

When the can is opened to start the measurements, radon gas diffuses through the filter at the top of the can. Once inside, a radon atom may radioactively decay. If the resulting alpha particle hits the piece of plastic in the cup (see Figure 8-5), it leaves a tiny "track," a microscopically small amount of damage in the plastic at the point where it hit. This process, where alpha particles hit the plastic, causing tell-tale "tracks," goes on until the can is sealed up again so that no more radon can get in.

The sealed-up alpha-track detector is then sent back to the laboratory from which it was brought. The processing technique used there is illustrated in Figure 8-5. The plastic is immersed in a caustic liquid chemical for a few hours. This chemical dissolves away the plastic around each of the alpha-particle damage sites, so that there is a hole or pit where each of the alpha particles hit the detector; an example is shown in Figure 8-6. From the number of holes, which can be counted under a microscope, the concentration of radon in the air can easily be estimated.

Like the charcoal-canister system, the alpha-track detector is inexpensive and simple. The plastic itself, which is actually the same as is used in many types of sunglasses, actually costs only a few cents. The cost is basically what the laboratory charges to count the pits in the plastic. The system is just as easy to use as the charcoal canister; again, it is simply a matter of opening up the container, leaving it — this time for some months — then closing it up and sending it off to the laboratory for processing.

In addition, a variety of specialized electronic devices can measure radon over short periods of time. The continuous radon monitor and the continuous working level monitor are devices that take radon or radon-daughter measurements in less than a day. Another technique is "grab sampling," where a small volume of air is "grabbed," or collected. This sample of air is then analyzed in a special detector (such as shown in Figure 8-7) to measure its radon or radon-daughter content. This measurement takes only a few minutes, which makes it highly affected by any variations in the local conditions. Two

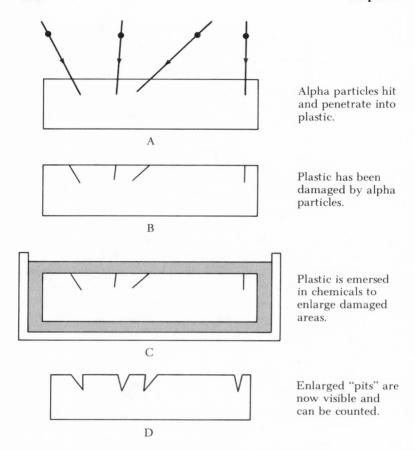

Alpha particles hit and penetrate into plastic.

A

Plastic has been damaged by alpha particles.

B

Plastic is emersed in chemicals to enlarge damaged areas.

C

Enlarged "pits" are now visible and can be counted.

D

FIGURE 8-5   How the alpha-track detector works. The top two pictures show alpha particles hitting the plastic and making microscopically small damage tracks in that plastic. The third picture shows the plastic being immersed in a caustic chemical to enlarge the damage sites. The final picture shows the plastic with enlarged "pits," which are now visible (see Figure 8-6) and can be counted.

readings taken with this type of instrument a few minutes apart are unlikely to produce the same answer. These measurements are expensive, but there may be occasions, such as the day of a real-estate closing, that require a same-day radon measurement. However, a glance at Figure 8-2 makes it clear that

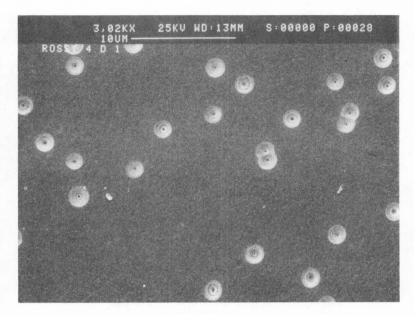

FIGURE 8-6   Enlarged photograph of a processed piece of sensitive plastic from an alpha-track detector. Each circle marks the place where one alpha particle struck the plastic.

measurements taken over short periods of time are not going to be representative of the average radon concentration in a house.

To summarize

Alpha-track detectors, which are inexpensive and easy to use, measure radon over a few months and thus are not very affected by short-term fluctuations in radon levels.

A variety of electronic radon detectors exist; they measure radon levels over short periods of time and can give very misleading results.

Both the charcoal-canister and alpha-track detectors are very reliable devices. They have no moving parts and essentially there is nothing to go wrong in them. The real difference

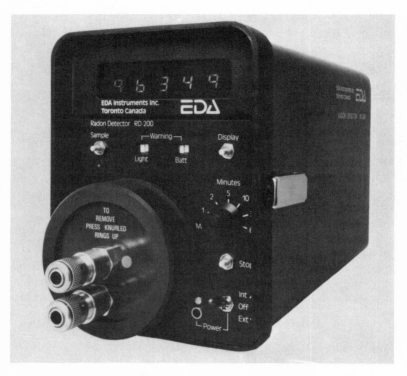

FIGURE 8-7 An electronic radon detector. This device works on the "grab-sampling" principle in that it draws some air into the detector and then analyzes it. Detectors such as this, which cost several thousand dollars, are not suitable for the long-term measurements most appropriate for screening or follow-up measurements. (Courtesy of EDA Instruments, Inc., Toronto.)

is that they measure different things. The charcoal canister measures radon levels averaged over a few days, whereas the alpha-track detector measures radon levels averaged over a few months.

So the most sensible plan of action is first to take a measurement with a charcoal canister for a few days. The result probably will be reliable to within a factor of about two: in other words, the true average radon level could be as much as twice as high or twice as low as the actual results of the measurement.

But this degree of accuracy is good enough for a first step. If, for example, the measurement result were 0.2 pCi/1, then it would hardly matter whether the real average is 0.2, 0.1, or 0.4: all would mean there is no further need for concern. At the other end of the scale, if the measured result were 50 pCi/1, then it really would not matter whether the real average was 50, 25, or 100 pCi/1: all would mean that something further needs to be done.

Let us now go step by step through the measuring procedure. We will consider:

- Where to get the detector
- How to use the detector
- What to do with the results of the measurement

WHERE TO BUY A DETECTOR

Having decided to find out the radon level in your home, the first step is to buy a detector. Currently, the easiest way to buy one is through the mail, though companies like Sears are planning to market detectors in their stores. As we have seen, the best detector to get for an initial measurement is the charcoal canister, so we will look at this device first.

The current type of charcoal-canister detector has been available only since 1985, yet well over a thousand companies in the United States are selling these devices. The basic idea is to buy (actually rent) a canister through the mail from one of the companies, open it up in the home and then return it to the company. A few days later the result will arrive through the mail. This procedure means that the homeowner is not seeing the actual processing of the canister at all. Fortunately, the Environmental Protection Agency (EPA) has set up a testing scheme called the Radon Measurement Proficiency (RMP) Program, in which it tests the reliability of each company's radon measurements.

The EPA holds these tests about twice a year and publishes a "Proficiency Report," listing the names of companies that have passed the test and such information as whether they have submitted an acceptable quality assurance plan. This report is available from the EPA (see Appendix A for details).

To try to reduce the 1,300 companies listed in the EPA report to a manageable number, let us imagine the ideal company. It would

- Have passed the EPA test every time it has been offered
- Have an appropriate quality assurance plan
- Have its own facilities for processing the radon detectors
- Be reasonably priced, say below $25

These requirements actually trim the field down to just two companies: they are listed in Appendix B. There are, of course, many other perfectly fine companies, some a little more or less expensive (the range appears to be from about $12 to $50). A long list can be found in the EPA report on its Radon Measurements Proficiency Program (see Appendix A). A few states also sell these detectors "at cost," which varies from $8 in New York to $18 in Maine. Details can be found in Appendix A.

As we shall see, if a second, follow-up measurement is needed, in most cases it should be done with an alpha-track detector. Again, many companies market these devices, but if we stick with the same standards that we used to pick charcoal-canister companies, we end up with just one company. Again, details are in Appendix B.

To summarize

There are over 1300 companies marketing radon detectors by mail.

The Environmental Protection Agency has a radon proficiency testing program to check the reliability of the different companies.

Some companies that are experienced and inexpensive are listed in Appendix B.

Having bought a detector, your next step is to expose it to the air in your house. Let us look at this step in more detail:

HOW TO USE A RADON DETECTOR

The first thing to decide about your new detector is where to put it. As we saw earlier, the nearer a room is to the ground, the higher the radon level. Consequently, a basement will have a higher concentration than a ground floor, which in turn will have a higher concentration than a second floor. A good rule of thumb is to put the detector in the lowest room in which anyone in the house spends any significant amount of time. The EPA actually suggests that the detector be put in the lowest room of the house, whether or not it is in use. This suggestion seems overly cautious, because the radon level in a room in which people rarely go is of little importance compared with that in more frequently used rooms.

The detector, when open, should not be in an enclosed place like a closet or drawer; it should be exposed to the same air as the people in the house are breathing. The best position is somewhere toward the middle of the room and a few feet off the ground — at breathing level. It should also be kept clear of sources of air turbulence, such as fans and windows. It should not be near a fire or radiator. A little care also should be taken to make sure children or pets do not move the detector.

The question of windows is important. As we have seen, if a window is opened, the radon concentration in the indoor air usually will go down, because it is being diluted by the outdoor air. To get the "worst-case" scenario, keep windows closed as much as possible during the time the detector is open. This can be quite impractical in the summer months, so it is much better to do the tests in the fall, winter, or spring. There is nothing wrong about doing the tests in the summer, but the results (see, for example, Figure 8-1) might be lower than those you would obtain if you took the measurements in another season — which might lead to a false sense of security.

Finally, when the charcoal canister has been open for two to seven days (the exact time is not very important, as long as it gets noted down on the form to be returned with the detector), it should be closed up and mailed back to the laboratory as soon as possible. Because the radon in the charcoal will be decaying away, the detector should be put in the mail *the same day* that it is closed up. The longer you wait, the less accurate the measurement will be.

To summarize

The radon detector should be placed in the lowest room in the house where any significant amount of time is spent.

It is important to keep windows closed as much as possible while the radon detector is open.

Testing during the summer is not recommended.

The detector should be placed toward the middle of the room, a few feet off the ground, and away from radiators, fires, fans, air conditioners, or windows.

The detector should be sent back to the laboratory for processing as soon as it is sealed up at the end of the testing period.

A few days after the detector is sent back to the company, they will send a letter back with the measured radon concentration.

WHAT TO DO WHEN THE RADON RESULTS COME BACK

The next step depends on the results of the first charcoal-canister measurement. As we saw in the last chapter, the aim of the exercise is to try to get the radon in the house below about 4 pCi/1. If the result of the first measurement is below that, by and large, you need do nothing more. However, if the measurement was taken in the summer, it could be unrealistically low; in this case, it would be a good idea to take another charcoal-canister measurement later, in winter.

If the reading was above 4 pCi/1 — the action level — further action is advisable. No matter how high the reading, the next step is to make a second measurement to confirm the first: it would be senseless to start making potentially expensive changes to the house without first double checking the measurement. What second measurement to take, however, depends on just how high the first measurement was. Basically, the higher the first reading, the greater is the urgency and the shorter the second measurement should be. In its "Citizens'

Guide to Radon" (see Appendix A), the EPA suggests the following reasonable scheme, summarized in Figure 8-8.

If the result of the first measurement was between 4 and 20 pCi/1, there is no enormous urgency. The best thing to do for the second measurement is to use an alpha-track detector and open it for one year. This long period will ensure that a realistic overall estimate of the radon levels in all seasons is obtained.

If the result of the first measurement was between 20 and 200 pCi/1, then there is a little more urgency. The follow-up measurement should also be done with an alpha-track detector, but this time it should be only for three months. As before, it is best not to do this measurement in summer, when radon levels are at their lowest.

Finally, if the result is more than 200 pCi/1 — which will be the case in only a very few houses — there would be no time

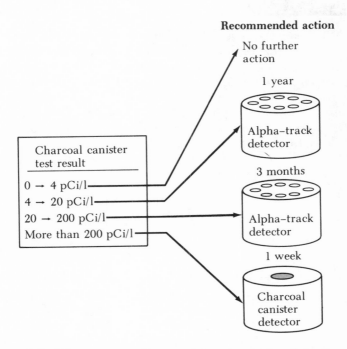

FIGURE 8-8 The next step after receiving the results of the first charcoal-canister test.

to lose. Another one-week charcoal-canister measurement should be taken, and then action would be urgently required.

To summarize

If the result of the first screening measurement is below 4 pCi/1, then there is normally no further need for action.

If the first screening measurement is between 4 pCi/1 and 20 pCi/1, then a follow-up one-year measurement should be made with an alpha-track detector.

If the first measurement is between 20 and 200 pCi/1, then a follow-up measurement should be made with an alpha-track detector for a three-month period.

If the first measurement is over 200 pCi/1, then a second charcoal-canister measurement should be taken immediately.

Whatever the level, if the second measurement confirms the first — so that both measurements have come out above the action level of 4 pCi/1 — the next step will be to consider some changes to the house to reduce the radon level. These measures can be as simple as opening windows or, more likely, will involve adding some plumbing to the house. The steps that need to be taken to diagnose and then fix a high-radon house are the subject of the next chapter.

How to Reduce
Home-Radon Levels

IF your follow-up test confirms that your house has a radon problem, then it is time to do something about it. If the house gets its water supply from a private well, a reasonable first step would be to check the water supply (see Chapter 10). If your water does not come from a private well, the radon is almost certainly coming from the soil and rocks under the house. Most people at this stage, unless they are expert at do-it-yourself, are going to turn to a contractor. We will go into detailed costing later; for now, the average cost to fix or mitigate the problem is roughly $1,000, consisting of $200 for diagnosing the problem and designing the solution, and $800 for actually doing the work. The actual price, which depends very much on the type of house and the amount of radon in it, could be as low as $400 or, rarely, as high as $3,000.

In this chapter we will go through the various stages of the mitigation process, from identification of radon-entry points through diagnostic testing to installation. It is not impossible for homeowners to do the work themselves, but unless they are prepared to do a good deal of research, it is probably not a good idea. At this time, radon reduction is almost as much an art as a science. The best techniques to use in a particular situation, as well as small but vital "tricks of the trade," are known primarily by trial and error from the many houses that have been worked on in the past. On the other hand, there are considerable cost advantages to "doing it yourself" because about three-quarters of a contractor's estimate will be for labor. So if you are determined, and want to reduce your bill from an average of $1,000 to just a few hundred dollars, then the best plan of attack is threefold:

- **Research:** read the EPA book *Radon Reduction Techniques for Detached Houses* (see Appendix A) from cover to cover.
- **Advice:** once you have specific plans, get several independent technical opinions about the plan in relation to your home.
- **Visit:** find a house that has had a similar mitigation system installed and examine it in detail.

If you are not planning to do the mitigation work yourself, the first step is to find a mitigator or mitigation company. This is obviously a vital step, so it is worthwhile spending some time on making the right choice.

The first problem is where to find lists of local mitigation companies. Often the best source is the state (see Appendix A), which in many cases provides lists of contractors, occasionally with some sort of state certification. If your state does not provide a list, local building or realtor associations are usually good sources. Some of the best sources of information, of course, are satisfied neighbors who have had mitigation work done.

Once you have located some mitigators, and are in the process of selecting one, it is important to have a checklist of questions to ask and points to be alert for when assessing them. A list of questions to ask would certainly include the following.

- Has anyone in the company taken a radon reduction training program, such as those offered by the EPA and by some states?

- How long has the company been doing radon mitigation? How many houses have they done? How many of them were structurally similar to yours?

- Will the company provide addresses of other houses on which it has worked? A list would enable you to talk directly to other homeowners who have used this company. Often the contractor will say that this information is confidential, but if they have done many houses, a few should certainly be willing to give references. Be very wary if they will give no references for other houses that they have mitigated.

- How much time will the contractor spend with you to discuss the various options? There will always be cost/benefit trade-offs: in other words, there is going to be a choice between more expensive systems that have a greater likelihood of a large radon reduction, and less expensive systems with a somewhat lower chance of a large reduction. The contractor should present and explain all the different options.

- Is the contractor suggesting a "phased approach"? In other words, to start with an inexpensive system and test the results: if that is satisfactory, then the job is finished; if not, then a second system can be installed. This approach is often very cost-effective.

- Will the contractor guarantee the cost of a proposed system? It is not reasonable to expect a guarantee that a given system will bring the radon level below some given value, such as 4 pCi/1. This is simply not predictable—beware if a contractor does guarantee this.

It is important to get more than one estimate: a reasonable aim would be to try and get three estimates, although four or five would be better. When comparing estimates, make sure that you are comparing estimates for the same type of system. In addition, find out if differing estimates include different numbers of diagnostic tests to help design the system. Finally, the cosmetic aspect is important. Some of the systems will involve a lot of extra piping. This can be concealed, though at extra cost. You, not the contractor, should decide what cosmetic items you want.

If you live in one of the big three radon states (New Jersey, New York, or Pennsylvania), a good idea is to call up the state radon officials (see Appendix A) and discuss the proposed plan with them. They are both knowledgeable and friendly.

One warning is in order, although this is hardly necessary for someone who has taken the trouble to read this book. As in any new field, there are some bogus mitigators. For example, a man in New Hampshire was selling elderly people a "radon remover" that had to be replaced once every month; it looked remarkably like a No-Pest strip!

To summarize

Radon mitigation of a typical high-radon house averages around $1,000.

An experienced and determined do-it-yourselfer could do the same job for around $250.

Assuming you do not want to do the job yourself, finding a good contractor to do the mitigation work is absolutely crucial and requires your time.

Many months can pass from the time that high radon levels are identified until the problem is actually fixed. If the level of radon in the house is particularly high, say more than 50 pCi/1, it is probably a good idea to take some temporary steps to reduce the radon levels right away. The best way to do this is simply to open windows. However, since this needs to be done all the time — or at least all the time that someone is in the house — it is not very convenient. On the other hand, it is only temporary.

There are two important points to note with this temporary ventilation system. First, the windows that are opened should be only in the lowest level of the house that has windows, so that the ventilation is as near the source of radon — the ground —as possible. If the house has a basement with windows, that is the room to ventilate; otherwise, ventilate the lowest living room. It is probably not a good idea to ventilate upstairs rooms, as this might create a flow of radon through the house and increase the amount of radon getting into the house. The second point to note is that more than one window in the room that is being ventilated must be opened: one open window could actually increase the radon level. The idea of opening more than one window is to create a flow of air across the room; this will both dilute the indoor air with low-radon outdoor air, and also reduce the pressure difference between the inside and the outside of the house. Both effects will reduce the indoor radon level.

This natural ventilation technique is quite effective and certainly costs nothing to install. Reduction of radon levels by a factor as great as ten (for example, from 40 pCi/1 to 4 pCi/1) can be achieved with this technique. On the other hand, in practice, open windows are inconvenient and could lead to security problems. In the summer and winter it will also be quite costly: it could double or triple heating or air-conditioning costs. One technique to limit the added energy costs is simply to open the windows only slightly, though less ventilation will result in smaller reduction in radon levels.

Let us now turn to more permanent measures of radon reduction. This aspect probably will be handed over to the mitigation contractor, but it is worthwhile understanding the three basic steps:

- Identification of radon-entry routes and causes
- Selection and installation of a radon reduction system
- Post-installation testing

IDENTIFICATION OF RADON-ENTRY POINTS

The first step in any radon mitigation scheme is to try to find the radon entry points. This task is by no means easy, because anything from large holes down to the tiniest of cracks is a possible route for radon to get from the ground into the house. While some openings may be easy to spot, others may be totally hidden or inaccessible. The inescapable fact is that houses are simply not built with airtight foundations.

Some of the major routes by which radon can enter a typical house are illustrated in Figure 9-1. Essentially, anything that creates a direct airway between the soil and the house is a possible pathway. Cracks in the concrete slab of a basement (route 1 in Figure 9-1) are common, as well as openings between poured concrete slabs and blocks (route 2). Sometimes a gap is deliberately left where the concrete basement floor meets the foundation wall (route 2); this drainage technique, usually called a french drain, is a likely entry route for radon. Other designs for draining basements involve floor drains or sumps (route 3), which frequently have exposed soil at the bottom, again providing a route for radon entry. Frequently,

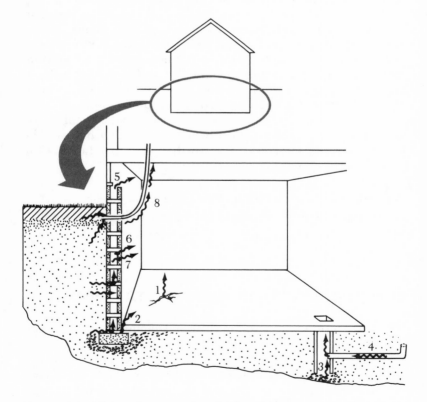

FIGURE 9-1   The main routes by which radon can enter a house. They are described in more detail in the text.

drain tiles from inside or around the house ("weeping" tiles) end up draining into the sump (route 4); in this way, radon from the area surrounding the house could be funneled into the house.

Another common radon-entry route is through hollow-block walls. Radon can get into the hollow parts of these walls through mortar joints, through cracks, or simply through the pores in the blocks themselves. (Cinderblock is particularly porous to radon gas.) Once inside the hollow walls, the radon can get into the house in any number of ways: through unsealed areas around windows and doors, through unclosed areas at the top of the wall (route 5), through mortar joints (route 6), or simply through the concrete block (route 7).

For one reason or another, many holes are invariably cut into the walls and floors of the basement. Gaps around the pipes through which utilities (water, sewer, gas, oil, and electricity) come into the house are very common. If these gaps are not perfectly sealed around the pipes, they can provide a significant entry route for radon (route 8 in Figure 9-1).

Another type of house design not illustrated in Figure 9-1 is one with a crawl space under the floor. The crawl space, often simply having a dirt base, is separated from the living area by a floor. Any gaps or cracks in this floor, such as for pipes, can be sources of radon entry.

With so many possible entry points to cover, the task is a bit like trying to block a river with a picket fence. Of course, it makes sense to block obviously large sources of entry, such as sumps (routes 3 and 4 in Figure 9-1), uncapped tops of block walls (route 5), or french drains (route 2). Sealing up these sources might not be beyond the capability of many enthusiastic homeowners. However, because it is rarely possible to find all the radon-entry points, this approach is not usually sufficient on its own to bring a high-radon-level house down below 4 pCi/1. Sealing off major entry points is therefore something which should be done in conjunction with more effective techniques, which we shall look at next.

To summarize

Every house, unless it is designed with radon in mind, will have many routes for radon gas to enter.

Cracks in the basement floor, gaps between wall and floor, sumps, block walls, and gaps around pipes are the most common entry points for radon.

No matter which other radon-reduction techniques are used, it is important to try and seal off as many radon entry points as possible.

THE MAIN RADON-REDUCTION TECHNIQUES

There are at least a dozen different approaches to the problem of reducing radon levels in houses, but three techniques, all based on the same general idea, are now by far the most common. This idea relates back to the discussion in Chapter 4 on the way that radon gets into the house. The reason it enters the house so efficiently is because of small air-pressure differences. If the air pressure indoors is just a tiny bit lower than the pressure outdoors, which can happen for a variety of reasons relating to wind or temperature, then radon in the soil will literally be sucked into the house. The trick, therefore, is to reverse this pressure difference by making the pressure lower outside the house than inside. If the pressure is lower outside than inside, then the radon simply will not be able to get into the house. The general idea is to use a fan to draw radon-rich air away from the region around the house, thereby lowering the pressure immediately outside the house. The three different techniques draw suction either through the drain-tile ventilation system, directly under the basement, or through block walls.

The only way to really judge the success rate of these different techniques is to see how they perform in real-life situations. So as we look at the different techniques, we will look at the results of a project to reduce radon levels in forty four houses in the Reading Prong region of eastern Pennsylvania. The project was conducted for the EPA by a commercial contractor (American Atcom Inc., Wilmington, DE) on a variety of homes whose radon levels varied from 6 to 1,200 pCi/1.

Let us now look at the three techniques and their advantages and disadvantages.

DRAIN-TILE VENTILATION

Many houses have perforated plastic or clay drain tiles running around all or part of the building. These tiles, often called "weeping tiles," are designed to collect water and drain it away from the foundations of the house. The water collected in the drain is routed either to an open point outside the house or to a sump in the basement, where it is pumped away. The drain tiles are rarely full of water; when they are not filled with water, they can contain radon-rich air. The idea is to attach a fan to the drainage system and suck air from it. This suction should reduce the pressure around the drain tiles, which are located near one of the main places where radon can enter the house, namely, the base of the walls of the house. If ground conditions are favorable, the suction may extend all the way underneath the house, providing a low-pressure region beneath the entire house.

If the drain-tile system discharges the water to a point outside the house, the usual technique is to attach a plastic pipe to this discharge line. A fan is then mounted on top of the pipe to pull air out of the drain system, as shown in Figure 9-2. To prevent the fan from sucking outside air from the point where the drain discharges water, a water trap is installed between the

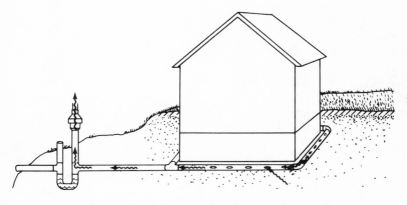

FIGURE 9-2   Illustrating the principle of drain-tile ventilation. The suction fan creates, through the drain tiles, a low-pressure region under the house.

fan and the outside air, as shown in Figure 9-2. If the drain-tile system discharges water instead to a sump in the basement (as in Figure 9-3), then the suction fan must be connected to the sump, which must also be made airtight.

The drain-tile ventilation technique is effective and fairly simple to install if drain tiles are already in place around the house. Sometimes the drain tiles do not go around all four sides of the house, but only around two or three sides. In this case it is still possible to use the technique, but the results probably will not be as good.

In fact, the drain-tile ventilation system is so simple that it is possible to "do it yourself." The cost of the materials (pipes, fittings, and fan) would be around $250. Bear in mind, however, that the major entry routes for radon must also be sealed off. If a contractor installs this system, the price would probably be in the region of $500 to $1,000. The cost would be some-

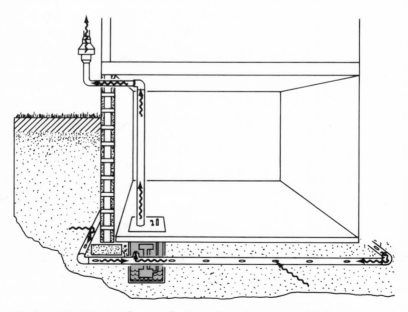

Figure 9-3 Drain-tile ventilation when the system discharges water to an internal sump. In this case, the top of the sump must be sealed and the fan connected to the inside of the sump.

what lower if the tiles drain into a sump inside the house. In this case, all that needs to be done is to put an airtight top on the sump and exhaust it to the outside with a fan.

As regards operating costs, the electricity to run the fan itself might cost about $30 per year. Because the ventilation in the house will be somewhat increased by this technique, heating and cooling bills could increase by about $100 per year.

Some results with this system from the test project in eastern Pennsylvania are shown in Figures 9-4 and 9-5. The results have been split up into the houses having drain tiles on all four sides, and those having drain tiles that do not completely surround the house. For houses having drain tiles on all sides, the average factor by which radon was reduced was about 30. (For example, if the original radon concentration was 100 pCi/1, then, on average, the system would bring it down to one-thirtieth of that, about 3 pCi/1.) Of course, in practice, there was a range of reduction factors, varying from about 4 in the worst case to about 80 in the best case. The results when the drain

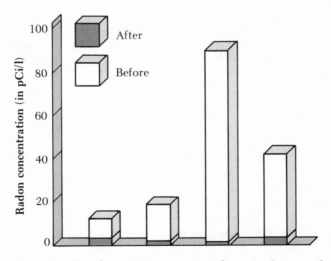

FIGURE 9-4   Results of an EPA test project for remediation of high-radon houses by drain-tile ventilation. In the four houses measured here, the drain tile completely surrounded the house. (Information derived from EPA reports EPA-600D 87-089/156, courtesy of D. B. Henschel and A. G. Scott.)

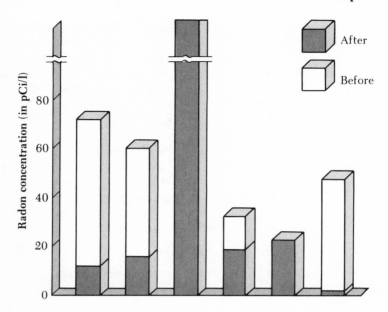

FIGURE 9-5 Results of EPA test project for six houses in which the drain tile only partially surrounded the house. Sizeable variation exists in the degree of success obtained, but, overall, the results are significantly worse than when the entire house is surrounded by a drain tile. (Information derived from EPA reports EPA-600D 87-089/156, courtesy of D. B. Henschel and A. G. Scott.)

tiles did not go around the house were not as good. The average reduction factor was about 7, with the best that was achieved in any house about 25.

If the drain tiles go around all four sides of the house, drain-tile ventilation is almost certainly the system of choice. It is relatively inexpensive and very convenient. If the drain tiles only go around part of the house, the decision is a little less clear-cut. The two factors then to consider are the radon concentration in the house and the soil permeability. If the radon level in the house is not too high—below 25 pCi/1, for instance—then drain-tile ventilation will probably be enough to bring the concentration below 4 pCi/1. The other factor is soil permeability. If the soil under the house is loose and permeable, the suction created around the drain tiles will be felt throughout the underside of the house, and the technique will probably work. On the other hand, if there is solid rock or hard-packed soil under the house, areas distant from the tiles may not feel the suction from the drain tiles, and the results will not be as good. Unfortunately, it is difficult to find out exactly what is under the house. One technique is to drill test holes in the basement floor and simply look down them; however, a lot of holes need to be drilled in order to find out the nature of the material everywhere under the house — which could leave the basement floor looking like a piece of Swiss cheese.

To summarize

If a house has drain tiles around all or three sides, suction on these tiles is usually the preferred radon-reduction technique.

The average cost of installing a drain-tile ventilation system is around $750 (less if the drain empties into an internal sump). Running costs will be about $100 per year.

The *average* radon reduction factor with this system is about 30 if the tiles surround all four walls; otherwise, around 7.

The success of the drain-tile ventilation system is usually related to the permeability of the material under the house.

SUBFLOOR SUCTION

If a house does not have drain tiles, the next option to consider is subfloor suction. In this technique (see Figure 9-6), plastic pipes are inserted into the dirt or soil beneath the concrete-slab basement floor. The pipes can be inserted either through holes cut in the concrete slab or from outside through the foundation walls. At the other end of the pipe, outside the house, a fan is installed to draw air away from the soil. The principle here is the same one that we saw for drain-tile suction: if the pressure is lowered under the basement floor, then radon-rich air beneath the house will not be able to flow through the floor into the house.

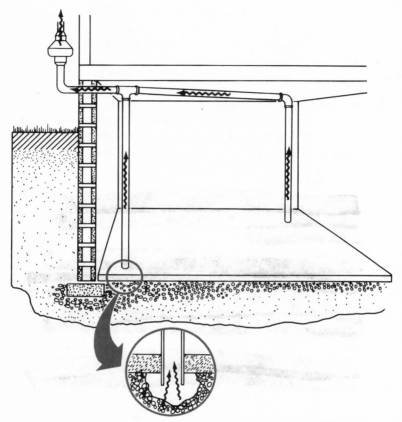

FIGURE 9-6   The principle of subfloor suction. Pipes, attached to a suction fan, are inserted into the ground below the basement floor, creating a low-pressure region under the house.

This technique is probably the most widely used of all. In about nine out of ten cases, it will reduce the house radon level below 4 pCi/1. The major complication is the same as with drain-tile suction: the permeability of the ground immediately under the house. If the ground is very impermeable, such as solid rock or tightly packed soil, then the suction from the pipe under the floor will be limited to a region around that pipe. What is needed, however, is for the suction to be felt over the whole area under the house. If there is a problem with impermeable soil, the solution is usually to install several pipes in different places going under the house; these can all be connected to the same suction fan.

Unfortunately, it is hard to predict how many pipes might be needed, because, as we saw with drain-tile suction, it is hard to know what is under the basement floor. The technique we discussed in that case was simply to drill holes and look down them. A more sophisticated way is to drill several holes through the basement floor, use a powerful vacuum cleaner to apply suction to one of them, and see whether there is significant pressure change at the other holes. This can be done, for example, by holding a cigarette or smoke stick near the other holes and watching the smoke. If the smoke is drawn into the hole, then the suction is effectively reaching that point.

In an average house, about one pipe is needed for every 700 square feet of floor area, which means that small, roughly square houses need only use one central pipe. Larger or odd-shaped houses may need two or even three pipes. If the ground permeability is poor, however, twice this number of pipes (or more) may be needed.

The cost of subfloor ventilation varies, particularly with the number and length of pipes that need to be installed. For a single-pipe system, the cost probably will be in the $500 to $1,000 range. If the pipe is taken up through the house and to the roof (rather than out of a nearby window), the cost would increase, with the benefit of not having an ugly fan at the side of the house. Running costs would be about the same as for the drain-tile ventilation system: about $30 per year to run the fan and about $100 per year in increased heating or cooling bills resulting from the extra ventilation caused by the system.

Finally, let us look at the performance of subfloor ventilation. The results from the eastern Pennsylvania study are shown in Figures 9-7 and 9-8 for 23 houses tested. The average reduction factor was about 40 (leading, for example, to a reduction from 200 pCi/1 to 5 pCi/1). There was, of course, a big variation from house to house in the reduction achieved. The best that was achieved was a reduction of over 150; the worst was a reduction of only 2.

For houses that do not have a concrete floor in the basement, but instead have an unpaved crawl space, a similar tech-

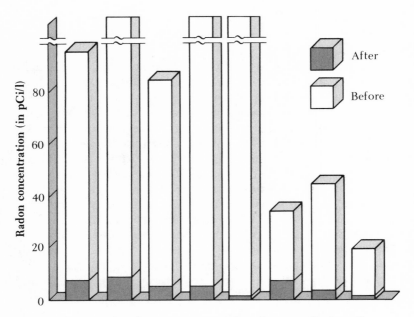

Figure 9-7   Results of an EPA test project for remediation of high-radon houses by subfloor suction. The eight houses measured here had concrete walls. (Information derived from EPA reports EPA-600D 87-089/156, courtesy of D. B. Henschel and A. G. Scott.)

nique is possible. In this case, a network of perforated pipes (like those used as drainage tiles) can be placed over the ground and attached to a suction pump. The pipes need to be covered by a roughly airtight barrier, such as a large plastic sheet, so that the fan can produce enough suction. The cost of this system is comparable to that of the more common subfloor ventilation system. Although it has not been used extensively, the performance of the crawl-space ventilation system should be as effective as subfloor ventilation.

To summarize

Subfloor suction is used in about three quarters of all radon-reduction projects.

The technique uses a fan to apply suction to the ground immediately under the basement floor.

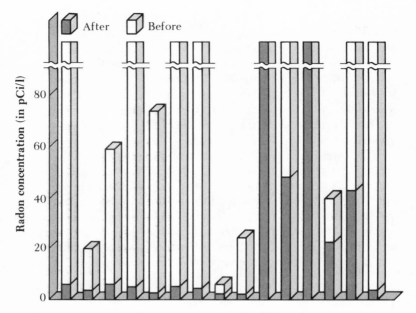

FIGURE 9-8 Results of EPA test project for fifteen houses with hollow, block walls. There is considerable variation in the degree of success obtained. (Information derived from EPA reports EPA-600D 87-089/156, courtesy of D. B. Henschel and A. G. Scott.)

The average installation cost is around $750 for a single pipe under the floor—and more if more pipes are needed; running costs are around $100 per year.

The average radon reduction factor for subfloor ventilation is around 40.

As with drain-tile ventilation, the success of the system is linked to the permeability of the material under the house.

BLOCK-WALL VENTILATION

The third option to reduce the air pressure around the house is block-wall ventilation. If a house has hollow block walls, a likely radon-entry route (see Figure 9-1) is either directly through the walls or through the joint of the wall and floor. In this case, radon-rich air builds up inside the hollow walls and can flow into the house through any opening in the wall. The reduction technique (Figure 9-9) is to put a pipe into each wall and connect them to an outdoor suction fan. The resulting suction in the wall cavity should make the air pressure in the wall cavity lower than the air pressure in the house. The air flow should then be from the inside of the house into the wall, preventing the radon-rich gas in the wall from flowing in the opposite direction, into the house.

The main problem with this technique is that if the walls are not airtight and are leaky, it is difficult to maintain effective suction. So it is very important to seal up all the major interior openings in the wall. Even if all the obvious openings are sealed up, it is hard to predict the effects of smaller openings, such as cracks in the mortar or even the porousness of the concrete block itself. The performance of the block-wall ventilation system is therefore rather unpredictable. For example, if there are

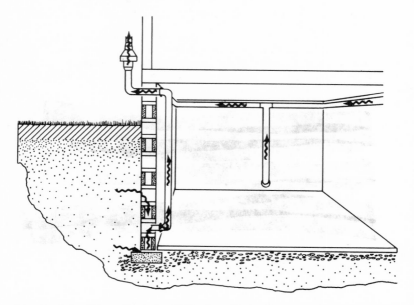

FIGURE 9-9   The principle of block-wall ventilation. Pipes, attached to a suction fan, are inserted in each wall, creating a low-pressure region around the house.

cracks remaining on the inside of the basement wall, the suction can actually cause a decrease in pressure in the whole of
the basement. If this happens, the radon level in the house
could actually increase, as radon-rich gas will be sucked out of
the ground and into the basement.

Instead of simply inserting pipes into the inside of the wall,
another (more expensive, less unsightly) option is to install a
"baseboard duct." Here (see Figure 9-10), a metal baseboard is
installed right around the room, covering the wall-floor joint.
Holes are punched at regular intervals in the part of the wall
covered by the baseboard. The whole region inside the baseboard is then attached to a suction fan. This system has the
advantage of also covering one of the major routes of radon
entry: the joint between the wall and the floor. On the other

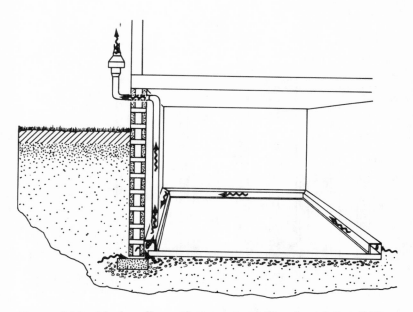

FIGURE 9-10   Block-wall ventilation using a "baseboard duct." The
same principle is used here as in the arrangement in Figure 9-9.
Here, however, the baseboard duct eliminates the need for most of
the unsightly pipes required in the more conventional
configuration of Figure 9-9.

hand, the baseboard-duct system is more expensive than simply inserting pipes in the wall.

The block-wall ventilation system is more expensive than either drain-tile or subfloor ventilation, because of the larger number of pipes (at least four, one for each wall) and because of the work required to seal off spaces and cracks in the walls. A typical cost would be around $1,500. The baseboard-duct system would cost about an additional $500 in extra installation work. Because of the increased flow of air through the walls, which are the main insulation for the house, the added cost in terms of extra heating and cooling bills may be significant: perhaps $300 per year for a single-fan system.

Some results for block-wall ventilation from the eastern Pennsylvania test project are shown in Figure 9-11. The aver-

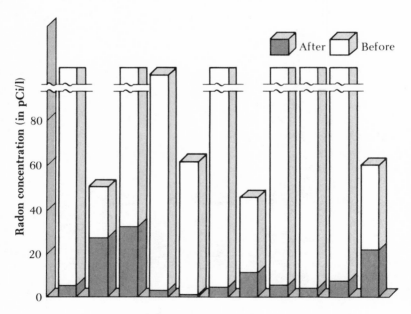

FIGURE 9-11   Results of EPA test project for remediation of high-radon houses by block-wall ventilation. Overall, the technique is less satisfactory than either drain-tile ventilation or subfloor suction. (Information derived from EPA reports EPA-600D 87-089/156, courtesy of D. B. Henschel and A. G. Scott.)

age radon reduction factor is about 30 (for example, reducing a 100 pCi/1 house to about 3 pCi/1), which is quite similar to the average reductions with the other two techniques.

Because the system is more expensive to install and operate than drain-tile or subfloor ventilation, it is the least desirable option. Its main application would be in houses with no drain-tile system and where the ground underneath is too impermeable for subfloor ventilation. In some houses with very high radon levels, both subfloor and wall ventilation have been tried together. In the eastern Pennsylvania test project, both techniques were tried on two houses: in the first house, subfloor ventilation brought the radon level down from 1205 pCi/1 to 520 pCi/1; adding wall ventilation brought the level down even further to 1 pCi/1. In the second house, subfloor ventilation brought the level down from 402 pCi/1 to 4 pCi/1, but adding wall ventilation actually *increased* the radon level back up to 26 pCi/1.

In conclusion, when we compare the three different mitigation systems, there is a very definite order of desirability. If there is a drain-tile drainage system going around two or more sides of the house, then drain-tile ventilation is probably the best system to use. If not and there is no indication that the house is built on rock or impermeable soil, then subfloor ventilation is the best choice. If this is not possible, then block-wall ventilation would be the option. Whichever system you choose, remember that a certain amount of sealing off of the major radon-entry points is always important.

To summarize

In block-wall ventilation, air is sucked from hollow basement walls to prevent a flow of radon into the house.

The results are less predictable than those with subfloor suction.

Typical installation costs are about $1,500 for pipes into all four walls and $2,000 for a baseboard-duct system. Running costs would be around $300 per year.

The average radon-reduction factor is around 30 but is rather unpredictable in individual cases.

In general, drain-tile ventilation is the first choice
between the systems, when available. If not, subfloor
ventilation is the best choice. Block-wall ventilation only
makes sense if the house is built on rock or impermeable
soil.

AIR CLEANING

Before moving on to post-mitigation testing, we should consider one other approach that has received a considerable
amount of publicity: air-cleaning. As we have seen, the radon
daughters that we inhale are usually attached to tiny pieces of
dust or aerosols. So it seems an attractive idea to use an indoor
air cleaner to remove them from the atmosphere, thereby making the air safe to breathe. There are, of course, many air
cleaners on the market.

To see why this technique usually does not work, we need
to refer back to Chapter 4, where we saw (Figure 4-17) that
although most aerosols are around four millionths of an inch
across, a few are much smaller, much less than one-millionth of
an inch across. When they are inhaled, these much smaller
particles are far more efficiently deposited on the lung than the
larger particles. Thus, radon daughters attached to small aerosols are more dangerous than radon daughters attached to large
aerosols. But what most air cleaners do is remove mainly large
aerosols from the air. What remains are mostly smaller-sized
aerosols. This means that radon daughters will be more likely to
attach themselves to small aerosols, which are comparatively
more dangerous. Thus, cleaning the air can actually be bad for
you!

The so-called electrostatic type of air cleaner does remove
some smaller particles from the air, and radon-removal devices
based on this type of air cleaner are on the market. However, at
best they will reduce the radon-daughter dose to the lungs by
factors in the region of 3 to 8, so they will not be adequate if the
indoor radon level is higher than about 30 pCi/1. In terms of
cost, at least one of these devices per floor would be needed,
each costing about $350. Running costs would probably be
low. There has been comparatively little testing of air cleaners
for radon reduction, so they are by and large an unknown

quantity. But it is certain that many air cleaners will actually increase the radon-daughter dose to the lung. So beware!

To summarize

Air cleaning produces a greater proportion of smaller-sized aerosols in the air. Radon daughters attaching to these smaller aerosols are potentially more dangerous than they are when attached to larger aerosols.

The best radon air cleaners will give a radon-dose reduction factor of about 5. Some cleaners will actually increase the dose to the lung.

Appropriate air cleaners cost about $350. Several would be needed in most houses.

POST-MITIGATION TESTING

Once the mitigation system is in place, both you and the contractor have to decide if it is working. As a first step, simply look at the system and check it over. Is all the planned piping installed? Are all the sections of piping connected to each other? Are the obvious openings and cracks filled up? The next step is to test the radon level to see whether it in fact has been reduced. The best way to do this is to use a charcoal-canister detector and take a three-day reading (see Chapter 8). As far as possible, conditions should be the same as when the first measurement was taken. If the conditions are not the same — for example, a window being open while it was previously closed — then the before/after comparison will be useless.

If contractors are doing the measurements, they probably will use an electronic continuous radon monitor. This device enables them to get continuous readings of the radon level over an extended period of time. There are two factors to be aware of here. First, as before, make sure that the windows are closed and that conditions in the house are as similar as possible to those when the original measurements were made. Second, the measurements need to be taken over at least six hours, to avoid measuring short-term fluctuation in radon levels. A good idea is to take a 24-hour reading with the system (that is, the suction fans) turned on, followed by a 24-hour reading with the system off. These measurements can also be done with charcoal canisters, with a three-day measurement time.

If the radon level is still not satisfactory, either the system was not installed correctly or the system is inherently not capable of adequately reducing the radon level. One way to make a rough check that the system is correctly installed is to check that suction is being produced. If the floor or wall is meant to be depressurized, hold a cigarette or a smoke stick near a crack or near a deliberately drilled small hole. If the suction is working, the smoke should be drawn into the wall or floor. If you do drill some small holes for this test, it is probably best to put them as far away as possible from the pipes providing the suction. In this way you can test the region where the least suction would be expected. This smoke test will not work for the drain-tile ventilation technique, as all the suction is produced underground.

If the system is providing some suction, yet the radon level is still too high, a more systematic series of measurements needs to be done. Pressure and flow measurements in the pipes can help to determine if there is a leak somewhere in the system or if a stronger fan is needed. In a subfloor ventilation system, pressure measurements under the floor can indicate whether more subfloor pipes are needed. Finally, radon measurements with a fast "grab sampler" (which will take a radon measurements in a few minutes) at various points throughout the house can be useful in identifying further points where radon is still entering the house.

To summarize

Radon testing should be done as soon as the mitigation system is in place.

Testing should be done under the same conditions that were in place during the original radon tests.

Testing should be done over at least twenty four-hours.

A quick way to see if the suction is reaching all over the floor and walls is to drill small holes and test the suction with a cigarette or smoke stick.

10
Radon in
the Water Supply

As we saw in Chapter 4, 400,000 to 800,000 homes in the United States may have an excessive amount of radon in their water supply. This means that 1 to 2 million people could be at risk. In this chapter we will look at the water problem in more detail: what it is, where it is, and what to do about it.

Household water comes from two different sources: underground water and rainwater from the surface. Surface water, which supplies about half the U.S. population, is virtually free of radon contamination. Underground water, however, will contain radon, because as this water flows around underground rocks, radon gas will dissolve in the water. The amount of radon that will dissolve in the water depends on the amount of radon (and therefore the amount of uranium) that is in the rocks.

When water is pumped from underground for domestic use, it will have radon gas dissolved in it. Mostly, this water is used in municipal or public water-supply systems. In this case, the water is usually stored for a few days while it is being treated for purification. This storage process not only releases radon from the water but allows most of the remaining radon to radioactively decay away. (Recall that the half-life of radon-222 is about four days.) Consequently, virtually all public water supplies have very low levels of radon, either because they use surface water or because the underground water has been treated and stored, removing the radon.

However, about one person in five in the United States gets water from an individual well or a small community system serving just a few homes. In these cases, the water is pumped directly from underground into the house, as needed. If radon has dissolved in the water, it will still be in the water when it is used in the home.

Drinking radon-rich water is not a problem. The alpha particles from the radioactive decays simply do not have the range to get out of the water and reach sensitive parts of the body. The problem arises when radon that is dissolved in the water is released into the air, increasing the radon concentration in the air. This release of radon gas from the water to the air happens whenever radon-rich water is exposed to air: it is called "de-gassing." The more the water is exposed to air, or aerated, the more radon will be released into the air. Thus, any devices that spray water, such as dishwashers, laundry machines, showers, or toilets, will be very effective at releasing radon into the air. In addition, hot water is more efficient in releasing radon than cold water, making dishwashers, laundry machines, and showers release even more radon into the air. Figure 10-1 shows the contributions of different activities to water-produced radon in the air. Washing clothes is the biggest contributor, not only because it involves agitation and spraying of hot water, but also because of the large amount of water used in an average wash.

The effect of water on indoor radon levels is dramatically illustrated in Figure 10-2. It shows the concentration of radon in the air in the bathroom of a house in Pennsylvania, in the same town where the Watras family lived. The normal radon level in the air was less than 2 pCi/1. Then a hot shower was run for about 15 minutes. The radon level in the air shot up to

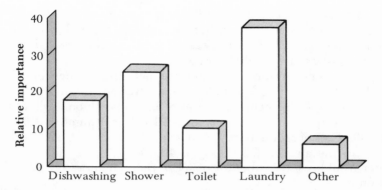

FIGURE 10-1   The relative importance of different everyday activities for the release of radon from water to the air. (Estimates are based on results from EPA reports EPA 600/2-78-173 and ORP/EERF-79-1, 1979.)

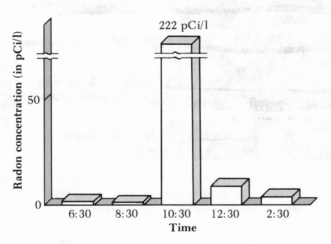

FIGURE 10-2   The rapid, brief rise in radon levels that can occur when water containing high concentrations of radon is used. In this case, radon levels are shown as measured in a bathroom before, during, and after a 15-minute shower. (Adapted with permission from M.C. Osborne, *Journal of the American Pollution Control Association*, 37. Copyright 1987, American Pollution Control Association.)

222 pCi/1, a more than hundredfold increase. Even four hours later, the bathroom still had not returned to the radon level before the shower. Similar increases have been observed when clothes washers are used in basements.

Radon concentrations in water are measured in picocuries per liter (pCi/1), just like radon concentrations in air. However, instead of a picocurie per liter of air, we are talking about a picocurie per liter of water. A good rule of thumb is that 10,000 pCi/1 of radon in water will contribute about 1 pCi/1 of radon to the air. So, for example, 40,000 pCi/1 of radon in water would result in about 4 pCi/1 of radon in air — the EPA action limit.

To summarize

High levels of radon in water are a potential problem only for the 20 percent of the population who use individual wells or small community water systems.

Drinking water containing levels of radon which occur naturally is basically harmless.

When radon-rich water is aerated (flowed through air), it releases the radon gas into the air.

Washing clothes, showering, and dishwashing are the daily activities that produce the biggest release of radon from water into air.

About 10,000 pCi/1 of radon in water will typically produce around 1 pCi/1 of radon in air.

Although all individual wells and small community water systems are suspect, some states tend to have especially high radon levels in private wells. No large-scale surveys have been carried out, but states that have a high average radon concentration in private well water include Florida, New Hampshire, Maine, Massachusetts, Pennsylvania and Rhode Island. Most studies have been in the state of Maine; a map showing the average concentration of waterborne radon in Maine is shown in Figure 10-3. As with all such geographic maps, they are only useful as average indicators and give no information about individual water supplies. For example, two houses in Maine no

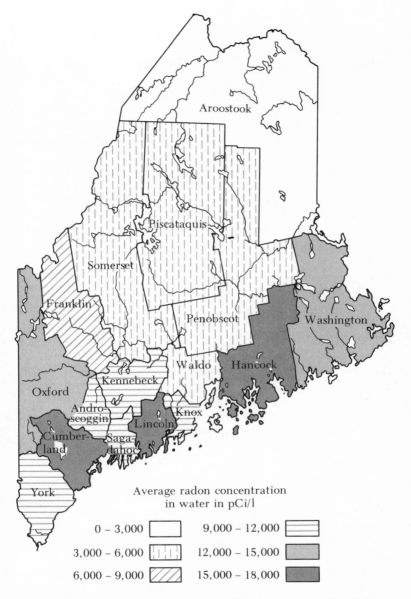

FIGURE 10-3 Average concentrations of radon in water for different counties in Maine. These results show average levels and do not, of course, give any information about water supplies in individual houses. (Reproduced with permission from C.T. Hess, et al., *Health Physics* **45**. Copyright 1983, Pergamon Press plc.)

more than 50 yards apart were reported to have radon levels in water of 1,250,000 pCi/1 and 16,000 pCi/1, respectively.

As of 1988, the EPA had not yet set any regulatory standards for radon in water, although the state of Maine has suggested an action level for water of 20,000 pCi/1. Of course if the radon level in the air is below 4 pCi/1, there is little point in taking any action. But if you have tested your house for radon in the air and the result was high, and your house water is supplied by a private well or small community system, then one of the prime suspects is water. The next step is to have the water tested for radon.

Household water is tested by putting a small amount (a little over a teaspoon) into two small bottles or vials and sending them off to a laboratory for analysis. Many of the companies that perform air-radon testing, including those listed in Appendix B, also perform water tests. The cost usually ranges from $15 to $50. Maine will also perform radon tests on water.

The testing company sends you the small plastic vials, which you then fill with water and send back for processing. The best time to take the sample is after a period of normal water use, such as after a morning shower. Some care must be taken when filling the vial, basically so that as little radon as possible will "de-gas" from the water as it is being transferred into the vial. If there is an aerator on the faucet, it should be removed. Then the cold-water tap should be run slowly for about 10 minutes. When putting the water in the vial, make every effort to avoid getting air bubbles in the water. This can be done by filling a cup with water, then withdrawing a small amount of water from the cup with a syringe (provided by the testing company) and finally injecting the water slowly into the vial. (The testing company will provide details of the exact procedure.) As soon as the vial is full, the cap should be replaced. The vials should be mailed back to the laboratory the same day.

The next step is to decide what to do on the basis of the results of the test. First, you need to compare the radon-in-air measurement, which should already have been done, with the radon-in-water measurement. The idea is to see whether the water could be the main source of the radon in the air. As we saw, 10,000 pCi/1 of radon in water will produce roughly 1 pCi/1 of radon in the air. Suppose the radon-in-air measurement is 30 pCi/l and the radon-in-water measurement is 10,000 pCi/l. Now the 10,000 pCi/l of radon in the water will contribute only about 1 pCi/l of radon to the air. Most of the 30 pCi/l of radon in the air is therefore from the ground. In this case, radon-reduction efforts should be concerned not with water, but with preventing radon from entering the house from the ground. On the other hand, if the radon-in-air measurement is 30 pCi/l and the radon-in-water measurement was 200,000 pCi/l, then 20 pCi/l, or most of the radon, may be from the water and radon-reduction efforts should be focussed on the water supply.

As with radon-in-air measurements, the results of water measurements can vary from day to day or from season to season. If the result of the test is high (say, more than 40,000 pCi/l), the next step is to do another test. If the result of the first measurement is below about 200,000 pCi/l of radon in water, a three-month test with an alpha-track detector is recommended for the follow-up measurement. This alpha-track detector comes in a small cup with a lead weight attached around the rim (see Figure 10-4). The entire system is placed upside down on the bottom of the water tank of a toilet. Every time the toilet is flushed, radon is released, which can be measured by the alpha-track detector. The company that markets alpha-track detectors for air measurements (see Appendix B) also markets the water alpha-track device.

If the radon level in the water is above 200,000 pCi/l, there is a little more urgency and a second measurement of water inside a vial is appropriate.

To summarize

If a house has a high indoor-radon level and is served by well water, it is a good idea to test the water to see if it is contributing to the problem.

Radon tests for water, as with air-radon tests, can be done by mail.

If the result of a radon-in-water test is between 40,000 pCi/l and 200,000 pCi/l, then a follow-up measurement should be taken by putting an alpha-track detector in the toilet water tank for three months.

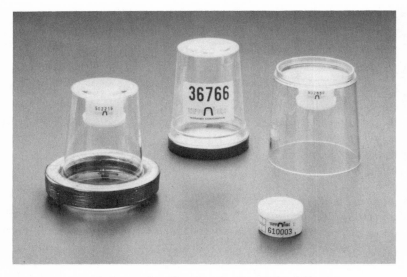

FIGURE 10-4   Alpha-track radon detector used for follow-up radon-in-water measurement. The detector is placed in a toilet water tank and held at the bottom by the lead weight around it. (Courtesy of Terradex Radon Detection Products, Illinois.)

If the result of radon-in-water test is above 200,000 pCi/1, then a second measurement made with water in a vial is the appropriate next step.

The techniques for removing radon from water are relatively simple. Flowing water over a material called granulated activated charcoal (GAC), which absorbs the radon in the water, is one technique which is generally reliable. (This charcoal is actually the same material as is used in charcoal-canister radon detectors.) The GAC unit, usually a fiberglass or metal tank, is installed in the main water-supply line from the well, after the pressure tank and sediment filter. The tank (see Figure 10-5), which is usually about four feet tall, is most often put in an out-of-the way place in the basement. The use of a sediment filter before the GAC unit is important to prevent the GAC unit from getting clogged up with dirt. Even with a filter, the GAC unit will need to be "backwashed" about once a year to remove dirt. This simply involves temporarily reversing the

FIGURE 10-5 A typical GAC (Granulated Activated Charcoal) unit—the narrow tank—for reduction of radon levels in water. The unit is in a basement where the water inlet for the house is located. In this photograph the GAC unit (left) is next to a conventional pressure tank. (Photograph courtesy of E.M. Moreau, State of Maine Department of Human Services.)

flow of water in the unit to flush out accumulated dirt. (In fact, it is a good idea to do this once when the system is first put in place.) Backwashing should not be done more than once a year because it temporarily reduces the efficiency of the unit; therefore, GAC units that automatically backwash themselves from time to time are not recommended.

GAC units come in various different sizes, containing between 1 and 50 cubic feet of activated charcoal. The larger the amount of activated charcoal, the more expensive the unit. If the level of radon in the water is below about 50,000 pCi/1, then the smallest GAC (1 cubic foot) would be adequate. To treat a water level in the 50,000 to 150,000 pCi/1 range, a 2-cubic-foot unit would be appropriate. For higher radon levels, a 3-cubic-foot unit should be used. Prices, including a sediment filter, range from about $800 for a small unit to about $1,200 for a 3-cubic-foot unit. Although no long-term tests have been carried out, these units probably last for several years.

One drawback to the GAC units is that the charcoal inside becomes mildly radioactive, because the absorbed radon decays to its various daughters (see Figure 4-1), which are themselves radioactive. Two of the daughters, lead-214 and bismuth-214, emit not only beta particles but also gamma rays. These gamma rays can penetrate through the walls of the GAC unit and are themselves potentially harmful. The higher the level of radon coming from the well, the higher the level of radioactivity coming out of the GAC unit. The nearer you are to the unit, the higher the radiation level, so one way around the problem is simply to cordon off an area of about ten feet around the unit. Generally, this solution is not very convenient; another approach is to immerse the GAC unit in a larger tank of water, about two to three feet across. The water in the tank would then absorb virtually all the gamma rays coming out of the unit. Such a tank would probably add about $300 to the cost of the system.

Because radon removal from water is a comparatively new art, there are currently few companies with expertise in this field. The best place to enquire is either with the state radon contact (see Appendix A) or with the state water authority. There are no certified companies in this area, though Maine has four companies on an informal list of qualified contractors.

Having installed the GAC, it is, of course, important to perform more radon measurements to see that the device is doing its job. Both a water-vial and an air charcoal-canister measurement should be taken. If the water level is reduced but the air-radon level is still too high, then there are significant sources of radon other than the water. In that case, the mitigation measures described in Chapter 9 will need to be considered.

Some time down the road you may want to discard the GAC unit. After it has been disconnected, it should be left for about a month, at which time most of the radioactivity will have decayed away. One exception is lead-210, which has a half-life of about 20 years. The activity of the lead-210 will be very low, but it probably would be a good idea to check with the state (see Appendix A) before discarding the GAC unit.

To summarize

The most common technique for removing radon from water supplies is to use granulated activated charcoal (GAC), which comes in a tank that is attached to the main water supply line.

A typical GAC system for removal of radon costs about $1,000.

The GAC unit will become, with use, mildly radioactive, emitting gamma rays. The best protective technique is to immerse the unit in a large tank of water.

After installation of a GAC, the radon level in both water and air should be retested.

It is advisable to check with the radiation authorities in your state (see Appendix A) before discarding a GAC unit.

Radon, Real Estate, and the Law

11

IN a recent survey, radon specialists were asked, "In your opinion, what is the motivating factor for homeowners requesting mitigation?" One in two said health fears were the main reason, but as many as one in three said real estate concerns were the most important factor. In the four years since the Watras house started the current level of concern over radon, there has been a surge of radon-related litigation concerning real estate transactions. As well as taking legal action against sellers and real estate agents for not telling them about radon, buyers are suing builders for faulty construction. In this chapter we will look at radon and real estate — where the responsibility lies, and who should do what during a real estate transaction.

Because the public concern is so recent, the legal ramifications of radon in real estate transactions are still not fully decided. In general, state law establishes the rules of the game; however, not only have many states not made decisions on radon issues, but those that have do not always agree with each other. What is clear is that the law treats radon gas exactly the same way it treats any other natural problem, such as termites, soil instability or flooding. Any legal rulings on problems like these will be directly applicable to the radon problem.

The question of who is responsible if a house is sold with some defect or other is an interesting one. Until the 1960s there was a universal doctrine called "caveat emptor," or "let the buyer beware." In the words of Milton Friedman, a Nobel Prize winner in economics, "If a purchaser signs a contract and then learns that the cellar collects water, the roof leaks, or the place is infested with termites, it is the purchaser's hard luck." This doctrine had held for many years, but things started to change after World War II. As mass-produced—sometimes poorly produced—houses hit the market, courts began to hold builders responsible for personal injury caused by a defectively built house. This change seemed only reasonable: unlike the builder, the potential buyer usually has no way of knowing all the possible defects in the house.

Today, the notion of "implied warranty" has replaced the doctrine of "let the buyer beware." An implied warranty says that the seller of a newly constructed home is a specialist and should be able to warranty that the home is built in a workman-like manner and is reasonably safe to live in. For example, a builder in New Jersey was sued for designing and installing a hot-water system that did not have a mixing valve, which would prevent very hot water from coming out. A child in the house was injured by the hot water, and the builder was held liable for negligence. The state Supreme Court ruled that the hot-water faucet was a dangerous but concealed feature of the house, which the builder but not the buyer should reasonably have known about. The analogy with radon is clear. There was a hidden defect (whether missing valve or radon) in the house about which the builder should reasonably have been aware, and thus should have done something.

This implied-warranty notion started in California but now has been adopted (see Table 11-1) in 37 states, with 10 states yet to decide. Only three states have stuck with the old buyer-

TABLE 11-1   *Position of the states on implied warranty*

Adopted implied warranty		Rejected implied warranty	Still to decide
Alabama	Nebraska	Georgia	Alaska
Arizona	New Hampshire	Tennessee	Delaware
Arkansas	New Jersey	Virginia	Hawaii
California	New York		Maine
Colorado	North Carolina		Massachusetts
Connecticut	North Dakota		Nevada
Florida	Ohio		New Mexico
Idaho	Oklahoma		Puerto Rico
Illinois	Oregon		West Virginia
Indiana	Pennsylvania		Wisconsin
Iowa	Rhode Island		
Kansas	South Carolina		
Kentucky	South Dakota		
Louisiana	Texas		
Maryland	Utah		
Michigan	Vermont		
Minnesota	Washington		
Mississippi	Wyoming		
Missouri	District of Columbia		

beware philosophy. Of the states that have accepted the idea of implied warranty, there is a split about whether only the first homeowner (i.e., the one who bought the house from the builder) can sue the builder or whether subsequent home-owners can also sue the builder. Many states (Arizona, Arkansas, Illinois, Indiana, New Jersey, Oklahoma, Texas, and Wyoming) have ruled that subsequent homeowners can sue the original builder. A few states (Connecticut, Mississippi, Montana, and South Dakota) have ruled that only the original houseowner can sue the builder. Ohio and Pennsylvania have ruled both ways!

There is a difference between high levels of radon and a faulty hot-water tap that injures a child. The long latency period between radon exposure and possible lung cancer makes radon, in most legal situations, a potential rather than an actual hazard. There will be no injuries to show the jury. However, in a recent case in South Dakota, the court ruled that people exposed to radon could sue for damages based on their *potentially* increased lung-cancer risk in the future, even though the disease had not actually appeared.

To summarize

Radon law is based on and similar to legal rulings for other natural problems in houses, such as termites or floods.

The old doctrine of "caveat emptor," or "let the buyer beware," has been replaced by the new concept of implied warranty, which emphasizes the seller's responsibility to sell a safe house.

If someone has been exposed to high radon levels, he may sue, even if no cancers have appeared, on the basis of a potential cancer.

Let us look at how these ideas actually impact on a real estate transaction. Suppose you buy a house that turns out to contain a high level of radon. It is important to know if anyone else can be held responsible for this state of affairs. Stanley Watras, whose house initiated all the current concern, complained, "Man didn't put [radon] in the ground; who are we going to sue, God?" As we have seen, however, although radon in the ground is unavoidable, radon in the home can be avoided with testing and appropriate remediation to a house. So at least three potentially responsible groups come to mind: builders, real estate agents, and the previous houseowners. Let us consider each of these in turn.

THE BUILDER'S RESPONSIBILITY

As we saw, a builder certainly has a responsibility, an implied warranty, to the buyer of a house that he builds. In many states, this responsibility will extend to subsequent purchasers of the house. If high radon levels are found in a house, there are two possible situations to assess in looking to see if the builder could be liable. In the first, a structural defect in the house causes a high level of radon; in the second, the house has a high radon level despite the fact that it is reasonably built. We will look at both possibilities.

Houses with Structural Defects

If a structural defect is the direct cause of high levels of radon in the house, then the builder is probably responsible. For example, a crack in the basement floor or poor sealing around pipes going through walls could be directly responsible for high radon levels. It probably does not matter whether the builder was aware that there could be a radon problem: he would have to accept responsibility because his work was not up to acceptable standards. The situation would be similar to cracks being found in the foundations of the house, resulting in flooding. In both cases poor workmanship has left the house open to a natural problem and caused the house to be dangerous to live in.

A famous (and ongoing) court case in this regard was brought by a Philadelphia physician, Joel Nobel. Nobel built a $300,000 energy-efficient underground house in Montgomery County, Pennsylvania and subsequently found it contained about 60 pCi/1 of radon. After more than six months of detective work and an expenditure of $100,000, the cause was traced to a leaky air-circulation system. Nobel sued the building contractor, demanding damages for the increased cancer risk faced by his family, repairs to his home, and the emotional distress and inconvenience caused by the problem. Not surprisingly, the case has been in the courts for more than five years and a final decision is still pending.

Houses Without Structural Defects

Suppose a contractor builds a house that does not have any obvious structural defects, yet still turns out to have a high radon level. Can the contractor be held liable? The answer is "yes" if the builder was aware of the problem but did not take steps to do something about it. For example, consider a contractor building a house on the Reading Prong, the region of Pennsylvania, New Jersey, and New York known to have a large risk of high radon levels. If the house was built in 1976, before the Watras discovery made radon infamous, it would have been quite understandable for the contractor to have built a house without giving a thought to radon. On the other hand, if he was building the same house in 1986, it would be unthinkable that he would not be aware of the radon problem, so he would almost certainly be liable. In the words of the Pennsylvania Supreme Court, the builder is in a far better position than the buyer to understand how to make a safe home; consequently, he should "bear the risk that a home which he has built will be functional and habitable in accordance with contemporary community standards." Of course, all houses contain some radon gas, so it would be reasonable that a builder would only be liable if the building had a radon level above the EPA action level of 4 pCi/1.

To summarize

If a house has high radon levels caused by structural defects such as cracks in the floor, the builder is probably liable.

If a house has high radon levels but no structural defects, the builder is probably only liable if the house was built recently in a high-risk area for radon.

REAL ESTATE AGENTS

The "middle man" in a house sale is usually a real estate broker. Let us suppose that through a real estate broker, someone buys a house that subsequently turns out to contain a high level of radon. What is the extent of the broker's liability?

There are several situations in which a broker could be at fault. The most obvious one is misrepresentation: here the broker knows of a radon problem in the house but does not tell the buyer. The situation is just the same as if the broker concealed any other defect, such as flooding or termites. For example, in 1984, a real estate broker was successfully sued for not disclosing that the basement of a building he had sold was subject to frequent flooding. As we saw, radon is considered just another hazard, like flooding.

Let us now suppose that the broker genuinely did not know that the house had a radon problem. If the house subsequently turns out to have high radon levels, is he or she then free from all liability? The answer, in general, is that the broker may still be responsible. The Code of Ethics of the National Association of Realtors says that a broker has a duty to actively seek out defects in a house that is being marketed. It says that a broker must not only "avoid concealment of pertinent facts" but "has an affirmative obligation to discover adverse factors that a reasonably competent and diligent investigation would disclose."

Just what is a "reasonably competent and diligent investigation"? It could mean that a broker must always perform an inspection, in this case for radon, and reveal the results to a buyer. Alternatively, it could mean that the broker is obligated only to explore in detail situations where there might be a reasonable basis for concern: for example, a broker in the Northwest might have no particular reason for believing that radon could be a local problem, whereas a broker in the Northeast might well have a strong basis for concern.

An example of how this works in practice is a recent case in California. A broker listed a house for sale that was built on earth that had not been adequately prepared and was prone to landslides. Although the sellers knew of these problems, they did not tell the broker. Soon after the house was sold, there were several landslides that damaged the house. The court ruled that, even though the brokers were not told of the problems, the brokers should have seen enough "red flags" — uneven floors, evidence of previous damage, and other clues — to have a reasonable suspicion of problems. They should then have investigated further themselves. Because they did not, the brokers were liable for damages. The situation is the same for radon. If it is reasonable to suspect high radon levels, based on other houses in the neighborhood, then the brokers have a duty to find out more about the radon situation in the house that they are marketing.

To summarize

If there is good reason for a realtor to expect radon in a house, the realtor is duty-bound to find out more about the radon levels in that house.

THE BUYER AND THE SELLER

Finally, we come to the main players in any real estate transaction, the buyer and the seller. If the sellers know that there is a high radon level in the house, they must tell the buyer. It would be plain fraud if they deliberately mislead the buyer, saying that there is not a problem. A more common situation is that the sellers know that there is a problem but simply do not reveal this fact to the buyers. In law, it is rare that someone can be sued for not saying anything, but this is not so in the case of dangerous defects in a house. Here, the courts have repeatedly found that the sellers must disclose any knowledge they have of dangerous problems in the home. For example, in 1974 the Pennsylvania Supreme Court held that a seller could be sued for not disclosing that the basement of a house was defective and had been flooded with sewage. The same court in 1982 held that a seller must disclose any previous termite infestation in the house.

For radon, there was a landmark case in 1981: the buyers of a house sued the seller who had not disclosed that the house was built on uranium-mill tailings (see Chapter 4), making the house very likely to have high indoor-radon levels. The Colorado Court of Appeals held that there were grounds for an action on the basis of fraud and deceit.

On the other hand, sellers who make a full disclosure about the results of any radon tests that have been done are almost certainly clear of any legal liability. This has led to the widespread use of radon-disclosure forms that are attached to home-sales contracts. Some examples, recommended by different boards of realtors, are shown in Figures 11-1 to 11-3. Typically, such forms do three things:

FIGURE 11-1 Sample Radon "Disclosure" form. Depending on the options checked, this form may allow any of the following options, assuming the house is tested and is above 4 pCi/l: seller does not want to remediate the house; the buyer may or may not be able to withdraw from the transaction; seller suggests a plan for fixing the house; or the buyer may or may not be able to withdraw from the transaction. (Form kindly provided by the Pennsylvania Association of Realtors.)

RADON DISCLOSURE ADDENDUM TO AGREEMENT OF SALE

_____ 19___

RE: PROPERTY _____
SELLERS: _____
BUYERS: _____
DATE OF AGREEMENT __ 19___, SETTLEMENT DATE __ 19___, SALES PRICE $__

1. BUYER acknowledges receipt of notice as set forth on reverse side hereof.

2. SELLER hereby acknowledges receipt of notice as set forth on the reverse side hereof, and certifies that:

( ) The property was tested and Radon was found to be at or below 0.02 working levels (4 picocuries/liter).

( ) The property was tested and Radon was found to be above 0.02 working levels (4 picocuries/liter).

   ( ) The property was modified after which it was retested and Radon was found to be at or below 0.02 working levels (4 picocuries/liter).

Seller does not warrant either the method or result of the test.

( ) I have no knowledge concerning the presence or absence of Radon.

3. BUYER'S OPTION (Check only one)

( ) BUYER acknowledges he has the right to have the buildings inspected to determine if Radon gas/daughters is present. BUYER waives this right and agrees to accept the property on the basis of SELLER'S certification and agrees to the release as set forth in paragraph 4 below.

( ) BUYER, at BUYER'S expense, shall within _____ days upon approval of this agreement, arrange a Radon test of the residential buildings on the property.

   If the inspection reveals the presence of Radon which exceeds 0.02 working levels (4 picocuries/liter), the BUYER within five (5) days of the receipt of the report shall furnish the SELLER with a copy of the test results. Upon receipt of the test results, the SELLER may within _____ days submit a corrective proposal, in writing, to the BUYER. Upon receipt of the corrective proposal, the BUYER shall, within five (5) days:

   a. Accept the proposal in writing, which action shall constitute a release as set forth in Paragraph 4 below; or

   b. Declare this agreement NULL and VOID, at which time all deposit monies paid on account shall be returned to the BUYER.

   Should the SELLER fail to submit a corrective proposal within _____ days, then the BUYER shall within five (5) days:

   a. Accept the property in writing, which action shall constitute a release as set forth in Paragraph 4 below; or

   b. Declare this agreement NULL and VOID at which time all deposit monies paid on account shall be returned to the BUYER.

   NOTE: There are various firms in Pennsylvania through which a Radon test can be arranged.

4. RELEASE—The BUYER hereby releases, quit claims and forever discharges SELLER, SELLER'S AGENTS, SUBAGENTS, EMPLOYEES and any OFFICER or PARTNER or any one of them and any other PERSON, FIRM or CORPORATION, who may be liable by or through them, from any and all claims, losses or demands, including personal injuries, and all of the consequences thereof, where now known or not, which may arise from the presence of Radon in any building on the property.

WITNESS _____ BUYER _____(s)
WITNESS _____ BUYER _____(s)
WITNESS _____ SELLER _____(s)
AGENT _____ SELLER _____(s)

RADON CONTINGENCY CLAUSE

The New Jersey Department of Environmental Protection (DEP) and the United States Environmental Protection Agency (EPA) have found elevated levels of naturally occurring radon gas in many areas of New Jersey. High levels of radon gas may effect the health and safety of residents occupying properties in these areas. The REALTOR and REALTOR-ASSOCIATES, who have participated in the sale of this property, are not qualified to evaluate all elements of this very complex problem and for this reason, they are unable to give advice to the Purchaser with regard to any concerns which the Purchaser may have about the possible presence of radon gas at the property. The Purchaser acknowledges that he/she has been advised by the REALTOR and REALTOR-ASSOCIATES who have participated in the sale of this property that it would be appropriate for the Purchaser to conduct his own investigation pertaining to existing radon-gas levels. In addition, the Purchaser acknowledges that he/she has been advised that certain radon-gas testing firms have equipment to detect elevated levels of naturally occurring radon gas on the property. Accordingly, the Seller and Purchaser agree that the Purchaser has the right, at his expense, to retain a qualified inspector to conduct and complete an investigation to test for elevated levels of naturally occurring radon gas on this property. Purchaser shall have ____( ) days from the date of this Agreement to have such radon tests conducted, using established testing procedures recommended by EPA and/or DEP. In the event that the test results indicate levels of radon gas above four (4) picocuries, Purchaser shall give Seller and REALTOR, within three (3) days of receipt of the test results, a copy of the test results together with an estimate from a qualified individual or company of the cost to lower these levels to the four (4) picocurie level. If the estimated cost for lowering the radon-gas levels is less than $____, Seller shall be obligated to perform the required work at Seller's expense. However, if the estimated cost of lowering the radon levels discovered exceeds $____, Seller shall, in writing and within five (5) days of receipt of the estimated costs, either (i) agree to give a credit to Purchaser at closing in the amount of the estimated mitigation costs, or (ii) notify Purchaser of its unwillingness to allow such a credit. In the event Seller shall be unwilling to allow such a credit, Purchaser shall, within five (5) calendar days of such notice, (a) waive the above contingency and be responsible for the cost of mitigating the radon-gas levels at his/her costs, or (b) void this Agreement in which event all deposit monies shall be returned to Purchaser and all further obligations between the parties shall terminate. In the event the Purchaser does not so notify the Seller within the above time periods, Purchaser waives his/her rights under this clause.

FIGURE 11-2 Radon contingency clause, which would be attached to a house-sale contract. According to the agreement described here, the purchaser has the right to withdraw from the transaction if the radon-test results proves to be high. (Form kindly provided by New Jersey Association of Realtors.)

RADON CONTINGENCY CLAUSE

The New Jersey Department of Environmental Protection (DEP) and the United States Environmental Protection Agency (EPA) have found elevated levels of naturally occurring radon gas in many areas of New Jersey. High levels of radon gas may effect the health and safety of residents occupying properties in these areas. The REALTOR and REALTOR-ASSOCIATES, who have participated in the sale of this property, are not qualified to evaluate all elements of this very complex problem and for this reason, they are unable to give advice to the Purchaser with regard to any concerns which the Purchaser may have about the possible presence of radon gas at the property. The Purchaser acknowledges that he/she has been advised by the REALTOR and REALTOR-ASSOCIATES who have participated in the sale of this property that it would be appropriate for the Purchaser to conduct his own investigation pertaining to existing radon-gas levels. In addition, the Purchaser acknowledges that he/she has been advised that certain radon-gas testing firms have equipment to detect elevated levels of naturally occurring radon gas on the property. Accordingly, the Seller and Purchaser agree that the Purchaser has the right, at his expense, to retain a qualified inspector to conduct and complete an investigation to test for elevated levels of naturally occurring radon gas on this property. Purchaser shall have _____ ( ) days from the date of this Agreement to have such radon tests conducted, using established testing protocols recommended by EPA and/or DEP. In the event that the test results indicate levels of radon gas above four (4) picocuries, Purchaser shall give Seller and REALTOR, within three (3) days of receipt of the test results, a copy of the test results together with an estimate from a qualified individual or company of the cost to lower these levels to the four (4) picocurie level. *Purchaser, at the time of delivery of the test results to Seller, shall determine whether he/she wishes to void this Agreement, in which event all deposit monies shall be returned to Purchaser and all further obligations between the parties shall terminate. If Purchaser chooses to go forward with this transaction, Seller shall be obligated to go forward with this transaction.* If the estimated cost for lowering the radon-gas levels is less than $_____, Seller shall be obligated to perform the required work at Seller's expense. However, if the estimated cost of lowering the radon levels discovered exceeds $_____, Seller shall, in writing and within five (5) days of receipt of the estimated costs, either (i) agree to give a credit to Purchaser at closing in the amount of the estimated mitigation costs, or (ii) notify Purchaser of its unwillingness to allow such a credit. In the event Seller shall be unwilling to allow such a credit, Purchaser shall, within five (5) calendar days of such notice, (a) waive the above contingency and be responsible for the cost of mitigating the radon-gas levels at his/her costs, or (b) void this Agreement in which event all deposit monies shall be returned to Purchaser and all further obligations between the parties shall terminate. In the event the Purchaser does not so notify the Seller within the above time periods, Purchaser waives his/her rights under this clause.

FIGURE 11-3 **Another possible radon contingency clause that could be attached to a house-sale contract. According to this agreement, if a radon-test result proves high, the seller must perform remedial work, assuming it costs less than some stated amount; if the seller does this remedial work, the buyer is committed to the transaction. (Form kindly provided by New Jersey Association of Realtors.)**

- They disclose the results of any radon tests that have already taken place in the house.
- They give the buyer the right to have the house tested for radon.
- Depending on the particular contract, if the radon-test result is high, they may give the buyer the option to walk away from the contract without penalty. Sometimes they specify that the buyer is still committed to the contract if the seller can mitigate the house so that the radon level is reduced below 4 pCi/1.

Unfortunately, there is enormous potential for fraud during a radon test conducted for real estate purposes. As we saw in Chapter 8, short-term (minutes or hours) radon measurements are extremely unreliable; the shortest reliable radon test is the charcoal-canister method, which takes at least two to three days. During this time, the seller is in possession of the house and the buyer is not. Anyone who has read this book could easily think of several ways for the seller to reduce the charcoal-canister reading. Opening windows would be one way, taking the canister upstairs would be another. Keeping it outdoors would guarantee a low reading. Of course, sellers who do tamper with a test in this way are liable for fraud once the buyers move in and conduct their own tests. Such fraud would be very hard to prove, however, because of the natural fluctuations in radon levels (see, for example, in Figure 8-2).

There are two ways to approach the potential problem of tampering with the radon detector. One is simply by negotiation: the seller agrees to reduce the price of the house by, say $1,500, an average cost of remediation, and the buyer agrees to take over all further responsibilities for testing and remediation after taking possession. The second approach is to set up an escrow account. Some amount—perhaps $3,000—is placed by the seller in an escrow account. After the buyers move into the house, they are required to conduct a long-term radon test, probably with an alpha-track detector. If the results require it, mitigation would then be done to bring the radon level down, paid for out of the escrow account. Any money in the account not spent would be split between the buyer and the seller, so that there is an incentive for the buyer not to overspend on the mitigation.

Both these strategies are in common use because they remove the incentives for a seller to tamper with a radon device. Another advantage is that the testing and remediation are not conducted under time pressure but can be done under optimal conditions.

To summarize

If the sellers know that a house has a high radon level, they must tell the buyers.

In high-risk radon areas, radon-contingency clauses are now routinely attached to home-sales contracts; they usually specify an immediate radon test in the house for sale.

Because the radon test should take several days, there is potential for tampering with the testing device to make the radon level appear lower than it really is.

To avoid the problem of tampering, either the sellers reduce the price and the buyers take over all further responsibility, or an escrow account is set up to pay for mitigation after the buyers have moved in.

Radon and Life

12

WHEN we face so many other life-threatening dangers, is radon really worth worrying about? Now that we have some idea of the specific risks of radon, we can look at them in the context of other risks, to see if radon is really significant compared with everything else. We will first consider radon in relation to the other sources of radiation to which we are normally exposed; then we will look more generally at radon in the context of all other life-threatening risks that we face in everyday life.

RADON AND OTHER SOURCES OF RADIATION

When most people think about the dangers of radiation, they do not think about radon, but rather about nuclear power plants, nuclear waste storage, and fallout from nuclear weapons tests. What are the different sources of radiation to which an average person in the United States is exposed? The answer is illustrated in Figure 12-1. What is shown is the average "effective equivalent" radiation dose to which an American is exposed. The term "effective equivalent" takes into account that some radiations are more harmful than others, and also that some parts of the body, such as the lung, are more sensitive than others to radiation. Figure 12-1 gives an indication of the relative dangers of these different sources of radiation.

The most obvious feature in Figure 12-1 is that radon almost totally swamps the other sources: it represents about 80 percent of the total. Of the other sources, medical irradiation resulting mostly from x-ray diagnostic procedures such as barium enemas or stomach examinations, with about 7 percent of the total, is next in significance. In a sense, of course, this is

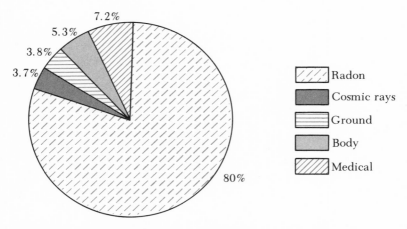

FIGURE 12-1   The relative amounts of radiation to which an average person in the U.S. is exposed from various different sources. (These percentage contributions to the average effective equivalent radiation dose were adapted, by permission, from the National Council on Radiation Protection and Measurements, Report No. 93, 1987. The figures were modified using recently published data by A.C. James in *Radon and Its Decay Products in Indoor Air*, W.W. Nazaroff and A.V. Nero, Jr., Editors. John Wiley and Sons, New York, 1988.)

misleading because not everyone has x-ray procedures, although a nationwide survey indicated that two out of three people in the United States have an x-ray examination each year.

The other main sources of radiation exposure are all naturally occurring and generally unavoidable. Radiation comes from the sky in the form of cosmic rays originating in and beyond our galaxy. It also comes from the ground, in the form of gamma rays from several naturally occurring radioactive materials in the ground. Finally, radiation comes from our own bodies; every cell in the body naturally contains tiny amounts of the radioactive atom potassium-40, which has a half-life of more than 1 billion years. When these potassium atoms radioactively decay, they produce beta and gamma rays, which result in a radiation dose throughout the body.

The more publicized radiation sources, from nuclear power production and from fallout from weapons tests, are not shown in Figure 12-1: they are so tiny in comparison with the sources that are shown in Figure 12-1 that they would be all but invisible on the figure. In fact, we are still receiving a small radiation dose from fallout from atmospheric nuclear-weapons testing. Fortunately, these tests had mostly ended by 1963, and since then the radiation dose caused by them has progressively decreased. It is now responsible for only about one-thousandth of the total radiation dose to which everyone is exposed.

Finally, we come to nuclear power plants. Nuclear power plants produce radiation exposure both to the plant workers and to the rest of the general public. The dose to the public comes primarily from nuclear waste, either in the ground or in the air. A 1982 United Nations report estimated that nuclear power production contributes about one part in 20,000 to the total "effective equivalent" radiation dose from all sources. Even if this figure underestimates the true dose from power plants by 10 or 100, it would still be tiny compared with the dose from radon.

These dose estimates for nuclear power plants are for normal operations, without accident. Figure 12-2 compares the average lifetime doses from radon exposure in the United States with the doses to which people were exposed as a result of the Chernobyl accident. (The Soviet disaster, which oc-

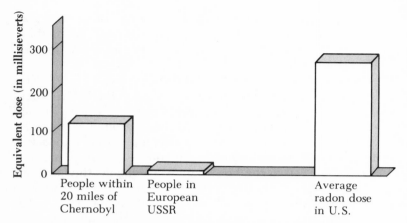

FIGURE 12-2 A comparison between the effective equivalent
radiation dose to which people are typically exposed over their
lifetime from radon and radiation doses from the Chernobyl
accident. Shown are the average equivalent doses to people living
within 20 miles of Chernobyl and were evacuated, and to the rest of
the population in the European U.S.S.R. Doses from Chernobyl to
people in the rest of Europe and to the United States were much
smaller. Doses from the Three Mile Island reactor accident were far
smaller than those from the Chernobyl accident. (Doses from the
Chernobyl accident from U.S. Department of Energy Report
DOE/ER-0332, 1987.)

curred in April 1986, was about as catastrophic an accident as
has happened to a nuclear power plant.) Because of radon, an
average person in the United States will be exposed, over his
lifetime, to more than twice the effective equivalent dose that
even the residents of Chernobyl received after the plant's mis-
hap. Closer to home, the worst nuclear-reactor accident in the
United States was at Three Mile Island, near Harrisburg, Pennsylvania. The largest "effective equivalent" dose to which any
member of the public was exposed because of this accident was
several hundred times less than the corresponding average lifetime dose from radon.

We can conclude that no other sources of radiation to
which we are exposed are remotely as important as radon. In
the next section, we will look at the risks of radon in the context
of all the other, non-radiation-related risks of dying in modern-
day life.

To summarize

Radon is by far the most important radiation hazard to which the public is exposed. It is responsible for about 80 percent of the total "effective equivalent" radiation dose.

This "effective equivalent" radon-daughter dose to which average people are exposed over their lifetimes is much bigger even than the dose to which people in the vicinity of Chernobyl or Three Mile Island were exposed.

RADON, LIFE, AND DEATH

"Why should I bother about radon, with all the other risks I face in my daily life?" This is a commonly asked and very relevant question. How does the risk of radon relate to all the other risks of everyday living? Is it worth worrying about? In this final section we will try to put the radon problem in some perspective.

There were about 240 million people living in the United States in 1988. Every year about 2 million Americans, or one in 120, die from some cause. Figure 12-3 shows the five main causes of death for Americans under the age of 70. By far the two largest causes are heart disease and cancer. After these two come car accidents, strokes, and, soon, AIDS. So, if we want to know the main risks to avoid, we need to concentrate on heart disease and cancer.

In terms of heart disease, there are certainly ways of reducing the risk of heart attacks. Heart disease is known to be strongly influenced by lifestyle. In particular, smoking, high blood pressure, and a high cholesterol level — as well as such other factors as being overweight and not exercising — increase the heart-disease risk. All these factors can to some extent be altered by changes in lifestyle.

The same is true, by and large, for cancer. There are about half a million deaths from cancer each year. The relative frequencies of the different types of cancer are shown in Figure 12-4. Lung cancer is by far the most frequent type of the disease both for men and women, being responsible for about 139,000 deaths per year. The next most common forms of the

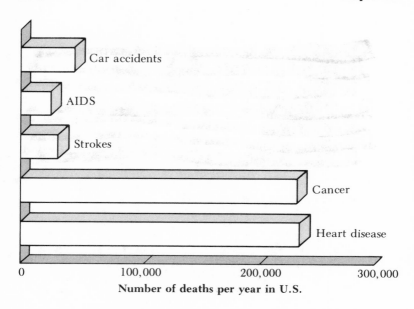

FIGURE 12-3 Numbers of deaths each year in the U.S. of people
under 70. Only the five largest "killers" have been included.
(Information based on "Vital Statistics of the United States," U.S.
Department of Health and Human Services, 1985. AIDS death rate
projected from number of AIDS carriers reported in the U.S.,
1987–1988, taken from "AIDS Weekly Surveillance Report,"
Center for Disease Control, Atlanta, 1988.)

malady are colon or rectum cancer and then breast cancer.
Although not definitely proven, it seems very likely that both
colon and breast cancer can be reduced by changes in lifestyle,
in particular, diet. Like heart disease, the risk of many common
cancers can be reduced by appropriate changes in lifestyle —
eating less fatty foods, for instance.

Let us turn to the most common cancer, and the one of
interest in the context of radon, lung cancer. The link between
lung cancer and smoking is well known. The lung-cancer death
rates for smokers and nonsmokers are compared in Figure
12-5. An average smoker has roughly a 10 times larger risk of
dying of lung cancer than a nonsmoker. Again, a change in

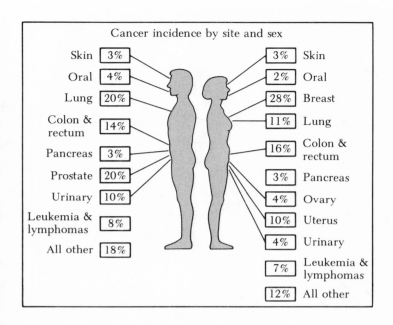

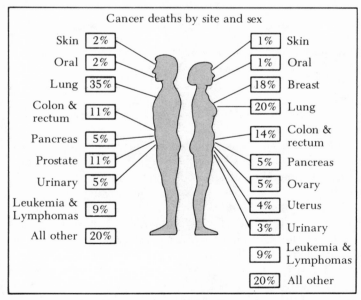

FIGURE 12-4 Cancer incidence and deaths according to type and sex. These estimates for these figures are for 1988. (From "Cancer Facts and Figures," copyright 1988, American Cancer Society.)

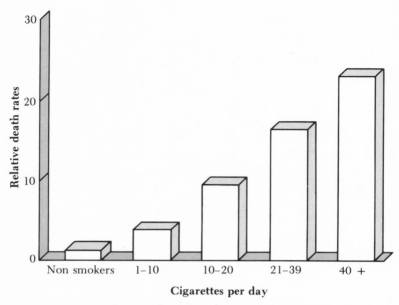

FIGURE 12-5   Death rates from lung cancer for smokers relative to
rates for nonsmokers. The conclusion is clear. (Results of a study of
U.S. Veterans adapted, with permission, from E. Rogot and J.L.
Murray, Public Health Reports 95, 1980.)

lifestyle — giving up smoking — can decrease a person's chance
of getting lung cancer.

Of course it is not easy to change lifestyle in accordance
with what is perceived as being healthy. In this regard, radon
holds a rather special position. Reduction of radon levels represents one of the few possibilities for decreasing one of the
major risks of dying *without* a significant change in lifestyle. As
we have seen, all that is required is a measurement, followed, if
necessary, by the remedial measures discussed in Chapter 9.
No difficult changes in lifestyle are required. The risks that can
be avoided may be considerable. For a person living in a house
with a radon level of about 4 pCi/1, the lifetime risk of dying of
radon-induced lung cancer is comparable to the risk of dying in
a motor-vehicle accident. A lifetime resident of a house that
contains 20 pCi/1 has a chance of dying because of radon that is

larger than that of dying from a stroke. If the house contains more than around 100 pCi/1, then radon-induced lung cancer will be the most likely way that its residents will die. The lifetime risk will be even greater than that of heart disease. As we have seen, these large risks can be drastically reduced without the need for a change in lifestyle.

Apart from reducing radon levels, there is another way to reduce the risk of living in a home with high levels of radon: stop smoking. Recall from Chapter 7 that, given the same amount of radon in a house, a smoker probably runs a much bigger risk than a nonsmoker of dying of radon-induced lung cancer. Stopping smoking will almost certainly reduce a person's risk not only of developing lung cancer from cigarettes, but also of getting lung cancer from radon. Giving up smoking *and* reducing radon levels is obviously the ideal.

To summarize

The two main causes of death in the United States are cancer and heart disease.

The risks of both heart disease and cancer are known to be reducible with changes in lifestyle, such as alterations in diet and smoking habits.

Reducing radon levels in the home is one of the few ways to reduce personal cancer risks without a change in lifestyle.

THE LONG-TERM SOLUTION TO THE RADON PROBLEM

For an individual living in a house with a high radon concentration, the right thing to do is to fix the house so that it has lower radon levels. The risk for the occupants of that house will then be considerably reduced. This solution is appropriate for affected individuals, but in fact such measures will not solve the overall radon problem for the population as a whole. To see this, consider Figure 12-6. This chart divides up the 16,000 radon deaths per year (projected in Chapter 7) between people

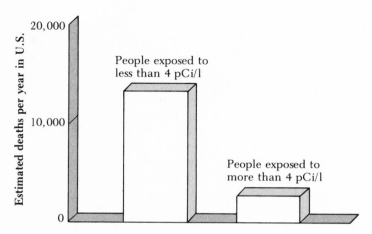

FIGURE 12-6   A division of the estimated 16,000 deaths each year
due to radon (see Figure 7-1) into numbers of people living in
houses with radon levels above and below 4 pCi/1.

who lived in houses containing 4 pCi/1 or less of radon, and
people who lived in houses containing more than 4 pCi/1. It
turns out that most of the deaths are of people who lived in
houses with low radon levels. This may at first seem paradoxi-
cal: how can there be more deaths of people exposed to low
levels than of people exposed to high levels, when we know
that the higher the radon level, the bigger the risk? The answer
is that there are far more people exposed to low concentrations
of radon than to high concentrations. Even though they run a
much smaller risk, there are so many more of them that they
end up showing more deaths than the high-dose group. In fact,
of the U.S. population of 240 million, about 20 million are in
the high-risk group (above 4 pCi/1), but 220 million are in the
low-risk group. So the conclusion is that there are more deaths
in the low-risk group.

It is not reasonable to expect people living in houses containing less than 4 pCi/1 to do anything about their radon
problem, because their personal risk is relatively small. Yet on a
national level, these low levels of radon are responsible for a
significant number of deaths. Here would seem to be the appropriate place for the government to enter the radon story. Its
aim would be to reduce the levels of radon in the entire
housing stock in the country. Clearly, it is not practical to

remediate every single house, but what the government can do is set rigorous building standards for *new* houses in terms of acceptable radon levels. Rather than make the universal action level 4 pCi/1, it would be quite feasible to mandate the action level for new houses to be 1 pCi/1. This would be relatively easy to achieve with a little extra attention to basement design and fabrication and with the more general use of drain-tile systems. The increase in costs of these new houses would be minimal. Several Scandinavian countries have already adopted the philosophy that new houses should have a more stringent standard than existing buildings.

The turnover rate for houses in the United States is relatively fast: about half the houses are replaced every 25 years. If building regulations were put in place now, by the middle of the next century most houses would have been built according to the more rigorous standards, and the overall average radon level in the United States would have significantly decreased. One state, Florida, recently tried to adopt a mandatory radon code for builders in "high-risk" areas, but the legislation has been stalled over the definition of a "high-risk" area.

In 1988, Congress passed a radon bill which states that the long-term goal of the United States is to reduce indoor radon levels to 0.2–0.5 pCi/1, the average outdoor reading. No direction, however, was given as to how to reach that goal, and the challenge has yet to be taken up by builders and legislators. If it is, future generations can expect to live in an environment where radon is a comparatively small problem. If nothing is done, the radon problem will not go away.

A final summary

Many more people are exposed to low radon levels than to high radon levels, which explains the fact that most radon-related deaths are of people exposed to low radon levels in their houses.

Deaths of people in low-radon level houses are, in the long run, mostly avoidable. What is urgently required is legislative action to reduce radon levels in new homes. Such action could reduce the death rate from radon to much lower levels within a few generations.

APPENDIX A

Where to Get Further Help

There are two sources of free help available to anyone who is worried about radon. The first resource is the Environmental Protection Agency (EPA), which can provide a number of free booklets on radon. The second resource is the individual states. A few states have very well-developed radon programs and can offer a great deal of help. Here are more details on how to get the most out of these two resources.

THE ENVIRONMENTAL PROTECTION AGENCY (EPA)

For the consumer, the EPA provides a series of booklets, running from very simple to quite technical, on the radon problem. They are listed here, with a short description of each, and details of where to get them.

A Citizen's Guide to Radon. What It Is and What To Do About It (August 1986, 13 pages). This very brief guide to the radon problem is designed for people who know nothing about it.

Radon Reference Manual (September 1987, 130 pages). An expanded version of the *Citizen's Guide.* A rather odd mixture of technical and non-technical material.

Radon Reduction Methods. A Homeowner's Guide (Second Edition, September 1987, 20 pages). A good, short guide to the different radon-reduction techniques. No detailed information.

Removal of Radon from Household Water (September 1987, 8 pages). Short introduction to the radon-in-water problem. No detailed information.

Radon Reduction Techniques for Detached Houses (Second Edition, January 1988, 208 pages). Extremely detailed description of the state of the art in radon reduction. The up-to-date second edition of this book is sometimes hard to get from EPA regional offices — try the EPA Center for Environmental Research Information, 26 West St. Clair Street, Cincinnati, OH 45268 (513-569-7931).

Radon Reduction in New Construction. An Interim Guide (August 1987, 9 pages). Brief description of techniques for keeping radon levels down in new houses, both before and after construction.

Radon/Radon Progeny Measurement Proficiency Program. Cumulative Proficiency Report (September 1987, 47 pages). List, the roughly 1,300 companies that have taken part in the EPA Radon Measurement Program. Describes which tests each company does and whether they passed or failed each of the EPA rounds of testing. Also divides the companies up by state.

The way to obtain any or all of these documents is to call your local EPA Regional Office. The EPA has divided the country into 10 regions. To find the right local office first consult Figure A-1. This will tell you what regional area your state is in.

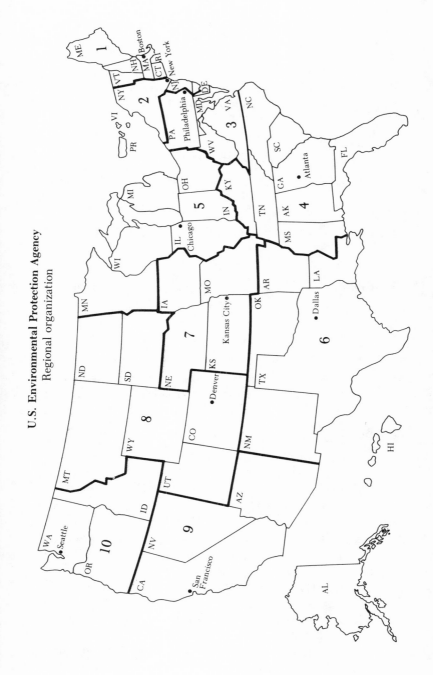

U.S. Environmental Protection Agency
Regional organization

Then, contact the appropriate regional office from the following.

EPA REGION 1   Room 2203, JFK Federal Building,
                      Boston, MA 02203.
                      (617) 565-3234.

EPA REGION 2   26 Federal Plaza, New York, NY 10278.
                      (212) 264-4418.

EPA REGION 3   841 Chestnut Street, Philadelphia, PA
                      19107.
                      (215) 597-4084.

EPA REGION 4   345 Courtland Street, NE, Atlanta, GA
                      30365.
                      (404) 347-2904.

EPA REGION 5   230 South Dearborn Street, Chicago,
                      IL 60604.
                      (312) 886-6175.

EPA REGION 6   1201 Elm Street, Dallas, TX 75270.
                      (214) 655-7208.

EPA REGION 7   726 Minnesota Avenue, Kansas City,
                      KS 66101.
                      (913) 236-2893.

EPA REGION 8   Suite 1300, One Denver Place, 999
                      18th Street, Denver, CO 80202.
                      (303) 293-1648.

EPA REGION 9   215 Fremont Street, San Francisco, CA
                      94105.
                      (415) 974-8378.

EPA REGION 10   1200 Sixth Avenue, Seattle, WA 98101.
                      (206) 442-7660.

STATE RADON PROGRAMS

The different states vary enormously in the amount of resources they devote to the radon problem. The states putting the most emphasis on radon are the ones that contain the Reading Prong: Pennsylvania, New Jersey, and New York. In addition, Florida (because of its phosphate lands) and Maine (because of its radon-in-water problem) have very active radon programs.

A recent survey of the states' responses to the radon problem divided the states into four groups, depending on how active they were in this field. A map showing which state is in which category is shown in Figure A-2. In the first category are states with "information programs." (About the limit of their involvement is to distribute EPA documents, which anyway can be obtained directly from the EPA. The second group of states have "formative programs." They are planning more active involvement, but as of 1988, from the consumer's standpoint, their potential help is also limited to distributing EPA documents. These states, however, are planning to do some statewide indoor radon testing to try to locate regions with elevated radon levels and, perhaps, "hot spots" with very high radon levels. The third group of states have "developing programs." As well as providing EPA literature, they have done or are doing extensive statewide indoor radon testing. So they do have, or soon will have, information on which areas of the state have a high likelihood of containing houses with high indoor radon levels.

The final group of states have "operational programs." They have a radon problem and they are actively addressing it. These states are Pennsylvania, New Jersey, New York, Maine, and Florida. They have done extensive radon testing, and many houses in each state have been remediated to reduce radon levels. They all have extensive information and training programs. Of these five operational-program states, the big three are Pennsylvania, New Jersey, and New York. They account for almost 90 percent of all the states' expenditures on radon and have more than half of all the state employees in the radon field. Table A-1 summarizes their consumer-assistance programs in a little more detail.

Finally, listed below are the agencies to contact for further radon information within every state.

Alabama   Radiological Health Branch, Alabama Department of Public Health, State Office Building, Montgomery, AL 36130.
(205) 261-5313.

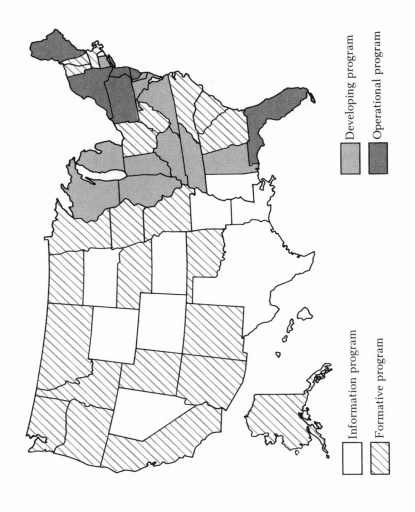

Developing program

Operational program

Information program

Formative program

TABLE A-1 Consumer-assistance programs of the "big three" radon states

State	Number of employees in radon program	Testing help for homeowners	Financial help for mitigation	Mitigation referrals	Number of phone enquiries
Pennsylvania	21	Free on Reading Prong	Low-interest loans	Provides list of state-certified companies	500–1000 per week
New York	19	Testing "at cost"; free with energy-conservation audit	Legislation pending on $400 subsidy for mitigation	Provides list of companies who have attended state training course; not state endorsed	900 per week
New Jersey	26	Free test if home is in known "hot spot" area; free second test for all homes where first test is above 4 pCi/1	Low-interest loans; legislation pending on tax deduction for mitigation	Provides list of companies in voluntary state-certification program; performance monitored through monthly reports and consumer complaints	1000–1500 per week

Alaska

Alaska Department of Health and Social Services, P.O. Box H-06F, Juneau, AK 99811-0613.
(907) 465-3019.

Arizona

Arizona Radiation Regulatory Agency, 4814 South 40th Street, Phoenix, AZ 85040.
(602) 255-4845.

Arkansas

Division of Radiation Control and Emergency Management, Arkansas Department of Health, 4815 W. Markham Street, Little Rock, AR 72205-3867.
(501) 661-2301.

California

California Department of Laboratories, 2151 Berkeley Way, Berkeley, CA 94704.
(415) 540-2469. (916) 445-0498 (Sacramento), (213) 744-3244 (Los Angeles).

Colorado

Radiation Control Division, Colorado Department of Health, 4210 East 11th Avenue, Denver, CO 80220.
(303) 331-4812/8480.

Connecticut

Connecticut Department of Health Services, Toxic Hazards Section, 150 Washington Street, Hartford, CT 06106.
(203) 566-8167.

Delaware

Division of Public Health; Delaware Bureau of Environmental Health, P.O. Box 637, Dover, DE 19903.
(302) 736-4731.

District of Columbia

DC Department of Consumer and Regulatory Affairs, 614 H Street, NW, Room 1014, Washington, DC 20001.
(202) 727-7728.

Florida

Florida Office of Radiation Control, Building 18, Sunland Center, P.O. Box 15490, Orlando, FL 32858.
(305) 297-2095, (305) 326-2095.

Georgia   Georgia Department of Natural Resources, Environmental Protection Division, 205 Butler Street SE, Floyd Towers East, Suite 1166, Atlanta, GA 30334.
(404) 656-6905, (800) 334-2373.

Hawaii   Environmental Protection and Health Services Division, Hawaii Department of Health, 591 Ala Moana Boulevard, Honolulu, HI 96813.
(808) 548-4383.

Idaho   Radiation Control Section, Idaho Department of Health and Welfare, Statehouse Mall, Boise, ID 83720.
(208) 334-5879.

Illinois   Illinois Department of Nuclear Safety, Office of Environmental Safety, 1035 Outer Park Drive, Springfield, IL 62704.
(217) 546-8100 or (800) 255-1245 (in state).

Indiana   Division of Industrial Hygiene and Radiological Health, Indiana State Board of Health, 1330 W. Michigan Street, P.O. Box 1964, Indianapolis, IN 46206-1964.
(317) 633-0153.

Iowa   Bureau of Environmental Health, Iowa Department of Public Health, Lucas State Office Building, Des Moines, IA 50319-0075.
(515) 281-7781.

Kansas   Kansas Department of Health and Environment, Forbes Field, Building 321, Topeka, KS 66620-0110.
(913) 296-1567.

Kentucky   Radiation Control Branch, Cabinet for Human Resources, 275 East Main Street, Frankfort, KY 40621.
(502) 564-3700.

Louisiana   Louisiana Nuclear Energy Division, P.O. Box 14690, Baton Rouge, LA 70898-4690.
(504) 925-4518.

Maine	Division of Health Engineering, Maine Department of Human Services, State House Station 10, Augusta, ME 04333. (207) 289-3826.
Maryland	Division of Radiation Control, Maryland Department of Health and Mental Hygiene, 201 W. Preston Street, Baltimore, MD 21201. (301) 333-3120 or (800) 872-3666.
Massachusetts	Radiation Control Program, Massachusetts Department of Public Health, 23 Service Center, Northampton, MA 01060. (413) 586-7525 or (617) 727-6214 (Boston).
Michigan	Michigan Department of Health, Division of Radiological Health, 3500 North Logan, P.O. Box 30035, Lansing, MI 48909. (517) 335–8190/8193.
Minnesota	Section of Radiation Control, Minnesota Department of Health, P.O. Box 9441, 717 SE Delaware Street, Minneapolis, MN 55440. (612) 623-5348 or (800) 652-9747.
Mississippi	Division of Radiological Health, Mississippi Department of Health, P.O. Box 1700, Jackson, MS 39215-1700. (601) 354-6657.
Missouri	Bureau of Radiological Health, Missouri Department of Health, 1730 E. Elm, P.O. Box 570, Jefferson City, MO 65102. (314) 751-6083.
Montana	Occupational Health Bureau, Montana Department of Health and Environmental Sciences, Cogswell Building A113, Helena, MT 59620. (406) 444-3671.
Nebraska	Division of Radiological Health, Nebraska Department of Health, 301 Centennial Mall South, P.O. Box 95007, Lincoln, NE 68509. (402) 471-2168.

Nevada

Radiological Health Section, Health Division, Nevada Department of Human Resources, 505 East King Street, Room 202, Carson City, NV 89710.
(702) 885-5394.

New Hampshire

New Hampshire Radiological Health Program, Health and Welfare Building, 6 Hazen Drive, Concord, NH 03301-6527.
(603) 271-4674.

New Jersey

New Jersey Department of Environmental Protection, 380 Scotch Road, CN-411, Trenton, NJ 08625.
(609) 530-4000/4001 or (800) 648-0394 (in state) or (201) 879-2062 (N. NJ Radon Field Office).

New Mexico

Surveillance Monitoring Section, New Mexico, Radiation Protection Bureau, P.O. Box 968, Santa Fe, NM 87504-0968.
(505) 827-2957.

New York

Bureau of Environmental Radiation Protection, New York State Health Department, Empire State Plaza, Corning Tower, Albany, NY 12237.
(518) 458-6451 or (800) 458-1158 (in state) or (800) 342-3722 (NY Energy Research and Development Authority).

North Carolina

Radiation Protection Section, North Carolina Department of Human Resources, 701 Barbour Drive, Raleigh, NC 27603-2008.
(919) 733-4283.

North Dakota

Division of Environmental Engineering, North Dakota Department of Health, Missouri Office Building, 1200 Missouri Avenue, Room 304, P.O. Box 5520, Bismarck, ND 58502-5520.
(701) 224-2348.

Ohio	Radiological Health Program, Ohio Department of Health, 1224 Kinnear Road, Columbus, OH 43212. (614) 644-2727 or (800) 523-4439 (in Ohio only).
Oklahoma	Radiation and Special Hazards Service, Oklahoma State Department of Health, P.O. Box 53551, Oklahoma City, OK 73512. (405) 271-5221.
Oregon	Oregon State Health Department, 1400 S.W. 5th Avenue, Portland, OR 97201. (503) 229-5797.
Pennsylvania	Radon Monitoring Program Office, PA-DER, Bureau of Radiation Protection, 1100 Grosser Road, Gilbertsville, PA 19525. (215) 369-3590 or (800) 23-RADON (in state only).
Puerto Rico	Puerto Rico Radiological Health Division, G.P.O. Call Box 70184, Rio Piedras, PR 00936. (809) 767-3563.
Rhode Island	Division of Occupational Health and Radiological Control, Rhode Island Department of Health, 206 Cannon Building, 75 Davis Street, Providence, RI 02908. (401) 277-2438.
South Carolina	Bureau of Radiological Health, South Carolina Department of Health and Environmental Control, 2600 Bull Street, Columbia, SC 29201. (803) 734-4700/4631.
South Dakota	Office of Air Quality and Solid Waste, South Dakota Department of Water and Natural Resources, Joe Foss Building, Room 217, 523 E. Capital, Pierre, SD 57501-3181. (605) 773-3153.

Tennessee   Division of Air Pollution Control, Custom House, 701 Broadway, Nashville, TN 37219-5403.
(615) 741-4634.

Texas   Bureau of Radiation Control, Texas Department of Health, 1100 West 49th Street, Austin, TX 78756-3189.
(512) 835-7000.

Utah   Bureau of Radiation Control, Utah State Department of Health, State Health Department Building, P.O. Box 16690, Salt Lake City, UT 84116-0690.
(801) 538-6734.

Vermont   Division of Occupational and Radiological Health, Vermont Department of Health, Administration Building, 10 Baldwin Street, Montpelier, VT 05602.
(802) 828-2886.

Virginia   Bureau of Radiological Health, Department of Health, 109 Governor Street, Richmond, VA 23219.
(804) 786-5932 or (800) 468-0138 (in state).

Washington   Environmental Protection Section, Washington Office of Radiation Protection, Thurston AirDustrial Center, Building 5, LE-13, Olympia, WA 98504.
(206) 753-5962.

West Virginia   Industrial Hygiene Division, West Virginia Department of Health, 151 11th Avenue, South Charleston, WV 25303.
(304) 348-3526/3427.

Wisconsin   Division of Health, Section of Radiation Protection, Wisconsin Department of Health and Social Services, 5708 Odana Road, Madison, WI 53719.
(608) 273-5180.

Wyoming   Radiological Health Services, Wyoming
                       Department of Health and Social Services,
                       Hathway Building, 4th Floor, Cheyenne, WY
                       82002-0710.
                       (307) 777-7956.

Radon Testing Companies

Listed here are some mail-order radon-testing companies that: are in the EPA Radon Measurement Proficiency Program; who have passed the EPA proficiency test every time it has been offered; that have an approved quality-assurance plan; that do not use another company to do the detector processing; and whose air detectors cost less than $25.

CHARCOAL-CANISTER DETECTORS

Key Technology, Inc: P.O. Box 562, Jonestown, PA 17038.
(717) 274-8310 or (800) 523-4964.
Detector Price: $19.99
(Water detector: $19.99)

TCS Industries, Inc: 3805 Paxton Street, Harrisburg, PA
                              17111.
                              (717) 657-7032.
                              Detector price: $18.55
                              (Water detector: $15.00)

ALPHA-TRACK DETECTORS

Terradex Corporation: 3 Science Road, Glenwood, IL 60425.
                              (312) 755-7911 or (800) 528-8327.
                              Detector price: $24.95
                              (Alpha-track water detector: $29.95)

Glossary

Absolute risk The risk of an adverse health effect that is independent of other causes of that same health effect.

Action level Concentration of radon in the house above which it is recommended that some action be taken to reduce that radon level; currently, the action level is 4 pCi/l in the United States.

Aerating Exposure to air of water that can cause radon dissolved in the water to be released into the air.

Aerosol Tiny atomized particles suspended in the air.

Air-exchange rate The rate at which indoor air is replaced with outdoor air; measures how airtight the house is.

ALARA "As Low As Reasonably Achievable": all levels of radiation are dangerous, so, ideally, all sources of radiation should be reduced as much as reasonably possible.

Alpha particle Atomic particle containing two neutrons and two protons that is ejected from a nucleus during the decay of some radioactive elements; for example, an alpha particle is emitted when either of the radon daughters polonium-218 or polonium-214 decays.

Alpha-track detector   A plastic detector used for measuring radon microscopically damaged when hit by an alpha particle. This damage is magnified by immersing the plastic in a liquid chemical, allowing the number of alpha particles that hit the plastic to be counted.

Atom   The basic building block of matter. Consists of a central nucleus with electrons orbiting around it.

Basal cells   Cells at the base of the wall of the lung airways. These cells divide to replenish the other cells in the lung wall and are often considered the key cells that, if damaged, can lead to lung cancer.

Becquerel (Bq)   Unit of radioactivity, corresponding to one radioactive disintegration per second.

Beta particle   Type of particle emitted during radioactive (beta) decay. A beta particle is actually a single electron, with either positive or negative charge.

Block-wall ventilation   Technique of radon reduction in houses in which the foundation walls are built with hollow concrete blocks or cinder blocks. Involves applying suction to the air spaces inside the wall.

Bronchi   The two main airways for conducting air from the trachea (windpipe) to the lung.

Cancer   An uncontrolled growth of cells that spreads indefinitely and increases in virulence.

Carcinogen   Any substance capable of causing cancer, such as radon daughters, cigarette smoke, or asbestos.

Cell   The basic unit of living matter.

Charcoal-canister detector   Used for measuring radon, it is a can of charcoal that adsorbs radon from the air. Subsequent analysis of the charcoal allows the concentration of radon in the air to be estimated.

Cilia   Hair-like cells that move back and forth in waves to propel foreign matter out of the lung and up into the throat.

Continuous-sampling detector A radon detector that takes a continuous reading of the radon level in the air. It is useful for revealing short-term fluctuations of radon levels, but less useful for estimating long-term average radon levels.

Control group A group of people subject to the same conditions as another group under study, except that the control group is not exposed to the specific factor being investigated in the study group.

Crawl space A non-living area beneath some houses, separating the floor from the ground.

Curie (Ci) Unit of radioactivity, corresponding to 37 billion radioactive disintegrations per second.

Daughter An atom formed as the result of a radioactive decay of a "parent" atom.

Decay series/chain A series of radioactive atoms, each the "daughter" of the one before and the "parent" of the one after; the series ends when any daughter is not radioactive.

Depressurization The process of reducing the pressure slightly, below or around the house, to prevent the flow of radon into the house.

DNA Deoxyribonucleic acid, a chemical usually found in the cell nucleus that contains the basic instruction code determining the function and replication of the cell.

Dose The amount of radiation energy that a given mass of material absorbs.

Drain-title ventilation One of the main techniques to reduce indoor radon levels. A fan is attached to an existing underground pipe system used for drainage, to reduce the pressure under the house.

Effective equivalent dose The radiation dose allowing for the fact that (a) some types of radiation are more damaging than others, and (b) some parts of the body are more sensitive to radiation than others.

Electrons Fundamental atomic particles that orbit around the atomic nucleus and occupy most of the space in an atom.

Epithelium A thin layer of cells that covers various organs in the body, such as the lung.

Follow-up measurement A second radon measurement, usually taken if the first radon measurement is high; its purpose is to make sure that the first measurement was reliable.

French drain A technique used for water drainage in basements: a small gap is left between the basement wall and the basement floor, to allow water to escape to the ground. A common entry point for radon.

Gamma rays High-energy, wave-like radiation emitted during the radioactive (gamma) decay of an atom.

Grab sampling A technique for measuring radon levels: it involves collecting a small sample of air into a container that can then be analyzed. Useful for quick measurements but gives no information about long-term average radon levels.

Granite Type of rock often associated with regions with high average radon levels in houses.

GAC Granulated Activated Carbon, a material that very efficiently removes radon from water flowing past it.

Gray (Gy) Unit of absorbed dose; equal to 100 rads.

Half-life The time it takes for one half of a group of radioactive atoms to undergo radioactive decay; each type of atom has its own characteristic half-life.

Initiating agent Something that causes initial "latent" damage to the DNA. The cell requires more damage from a second "promoting agent" before the damage is expressed as cancer. Radiation is usually considered an initiating agent.

Ionizing radiation Radiation that is harmful because it can knock electrons out of atoms in and near DNA.

Latency period The time between an injury occurring and the effects of the injury expressing themselves as disease. For radon-induced lung cancer, the period between exposure to radon and the appearance of lung cancer averages about 20 years.

Leukemia A cancer of certain blood cells; previously thought to be the main effect of radiation, but this is no longer believed to be true.

Lifetime risk   The risk of dying of some particular cause over the whole of a person's life.

Mitigation   The process of reducing indoor radon levels.

Mucus   Protective, sticky coating that lines, for example, the upper part of the lung; responsible for trapping foreign matter, which is then removed from the lung by the cilia.

Mutation   An abnormal change in the DNA of a cell or cells. Depending on the cell affected, it could lead to effects in that person or to effects in that person's offspring.

Neutron   One of the two particles that, with the proton, make up the atomic nucleus. The basic difference between the neutron and the proton is that the proton is electrically charged, whereas the neutron is not.

Nucleus (of atom)   The central part of the atom. The nucleus takes up only a tiny fraction of the volume of an atom but contains almost all of the atom's weight.

Nucleus (of cell)   The central part of a cell containing the genetic information bound in DNA.

Parent   Radioactive atom that disintegrates to give a different atom, its daughter.

Permeability   A measure of how easily gas and air can flow through a given material. Highly permeable materials under the basement are very desirable for drain-tile or subfloor suction systems.

Picocuries per liter (pCi/1)   A unit of measurement of the activity concentration of a radioactive material; measures, for example, how many radioactive disintegrations of radon occur every second in a liter of air.

Porosity   A description of a material relating to the amount and size of its internal pores.

Precursor   A precursor of an atom in a radioactive decay chain is a member of the decay chain that occurs before the atom in question.

Primordial   Existing at the beginning of the universe, or at the beginning of the earth.

Progeny   A daughter in a radioactive decay chain.

Promoting agent Something that acts on earlier cellular damage caused by an "initiating agent"; can cause the earlier damage to be expressed as cancer. Tobacco smoke is usually considered a promoting agent.

Proton One of the two particles that, with the neutron, make up the atomic nucleus.

Rad Unit of radiation dose. One rad is one hundredth of a gray.

Radioactivity The spontaneous emission of radiation by certain (radioactive) atoms, resulting in the transformation of one atom into a different one.

Radon daughter Any atom that is below radon-222 in the uranium decay chain; often specifically refers to polonium-218 and polonium-214, as these have the most biological significance; sometimes referred to as radon progeny.

Reading Prong Geological region extending from Reading, Pennsylvania, at the southern end, through Pennsylvania, New Jersey, and New York. Houses built on the Reading Prong have a higher-than-average chance of having high radon levels.

Relative risk Situation when the risk of a disease resulting from some injury is expressed as some percentage increase of the normal rate of occurrence of that disease; in contrast to an absolute risk, where the risk of a disease resulting from an injury does not depend on the normal rate of occurrence of that disease.

Rem Unit of equivalent or effective equivalent dose; one rem is one-hundredth of a sievert.

Remediation The fixing of a home to reduce its indoor radon level.

Screening measurement First measurement, usually done with a charcoal canister, of radon levels in a house.

Sievert (Sv) Unit of equivalent dose or effective-equivalent dose. One sievert is one-hundred rem.

Smoke stick A tube that releases a stream of dense, inert gas when a rubber bulb at the end is squeezed; useful for ascertaining details of air flow during remediation.

Stack effect Upward movement of air through a house, usually when the weather is cold outside; tends to force potentially radon-rich gas from the soil into the house, to compensate for indoor air leaking out of the top of the house. The name is derived from the similar flow of hot air up a chimney stack.

Subfloor ventilation Sometimes called subslab ventilation; the most common form of remediation for high radon levels, involving applying suction through pipes to the region underneath the basement floor.

Sump A hole in the basement floor designed to be a water drain; sometimes contains a pump to move the collected water out of the house.

Synergism The combined effects of two hazardous substances being larger than the sum of the separate effects of the individual substances.

Thorium series Radioactive-decay chain starting with thorium-232; one member of the chain is radon-220. This chain is of much less significance than the uranium-decay chain containing radon-222.

Threshold A level (for example, of radiation dose) below which there is no observable effect; there is no threshold for induction of cancer by radiation: all levels of radiation are harmful.

Trachea The windpipe through which air passes on its way to the bronchi and the rest of the lungs.

Tumor A proliferation of cells that do not have the normal mechanisms for controlling their growth.

Unattached fraction That fraction of the radon daughters in the air that are attached to extremely small particles. If inhaled, they have a greatly increased chance of being trapped on the wall of the lung.

Uranium series Radioactive-decay chain starting with uranium-238 and containing radon-222 and its daughters.

Weeping-tile Groundwater drain pipe that can be used for drain-tile ventilation.

Working level (WL) A measure of the total concentration of radon daughters that emit alpha particles.

Working level month (WLM) A measure of a person's exposure over time to radon daughters.

Index

Note: Page numbers in *italics* indicate illustrations; those followed by t indicate tables.